Solution Synthesis, Processing, and Applications of Semiconducting Nanomaterials

Solution Synthesis, Processing, and Applications of Semiconducting Nanomaterials

Special Issue Editor
Julia W. P. Hsu

MDPI • Basel • Beijing • Wuhan • Barcelona • Belgrade • Manchester • Tokyo • Cluj • Tianjin

Special Issue Editor
Julia W. P. Hsu
University of Texas at Dallas
USA

Editorial Office
MDPI
St. Alban-Anlage 66
4052 Basel, Switzerland

This is a reprint of articles from the Special Issue published online in the open access journal *Nanomaterials* (ISSN 2079-4991) (available at: https://www.mdpi.com/journal/nanomaterials/special_issues/nano_solution).

For citation purposes, cite each article independently as indicated on the article page online and as indicated below:

LastName, A.A.; LastName, B.B.; LastName, C.C. Article Title. *Journal Name* **Year**, *Article Number*, Page Range.

ISBN 978-3-03928-402-3 (Pbk)
ISBN 978-3-03928-403-0 (PDF)

© 2020 by the authors. Articles in this book are Open Access and distributed under the Creative Commons Attribution (CC BY) license, which allows users to download, copy and build upon published articles, as long as the author and publisher are properly credited, which ensures maximum dissemination and a wider impact of our publications.

The book as a whole is distributed by MDPI under the terms and conditions of the Creative Commons license CC BY-NC-ND.

Contents

About the Special Issue Editor vii

Julia W. P. Hsu
Solution Synthesis, Processing, and Applications of Semiconducting Nanomaterials
Reprinted from: *Nanomaterials* **2019**, *9*, 1442, doi:10.3390/nano9101442 1

Xianfeng Zhang, Engang Fu, Yuehui Wang and Cheng Zhang
Fabrication of Cu_2ZnSnS_4 (CZTS) Nanoparticle Inks for Growth of CZTS Films for Solar Cells
Reprinted from: *Nanomaterials* **2019**, *9*, 336, doi:10.3390/nano9030336 5

Takashi Nakamura, Hea Jeong Cheong, Masahiko Takamura, Manabu Yoshida and Sei Uemura
Suitability of Copper Nitride as a Wiring Ink Sintered by Low-Energy Intense Pulsed Light Irradiation
Reprinted from: *Nanomaterials* **2018**, *8*, 617, doi:10.3390/nano8080617 15

Dongyue Jiang, Yu Zhang, Yingrui Sui, Wenjie He, Zhanwu Wang, Lili Yang, Fengyou Wang and Bin Yao
Investigation on the Selenization Treatment of Kesterite $Cu_2Mg_{0.2}Zn_{0.8}Sn(S,Se)_4$ Films for Solar Cell
Reprinted from: *Nanomaterials* **2019**, *9*, 946, doi:10.3390/nano9070946 29

Hongseok Yun and Taejong Paik
Colloidal Self-Assembly of Inorganic Nanocrystals into Superlattice Thin-Films and Multiscale Nanostructures
Reprinted from: *Nanomaterials* **2019**, *9*, 1243, doi:10.3390/nano9091243 45

Hsin-Jung Wu, Yu-Jui Fan, Sheng-Siang Wang, Subramanian Sakthinathan, Te-Wei Chiu, Shao-Sian Li and Joon-Hyeong Park
Preparation of $CuCrO_2$ Hollow Nanotubes from an Electrospun Al_2O_3 Template
Reprinted from: *Nanomaterials* **2019**, *9*, 1252, doi:10.3390/nano9091252 61

Boya Zhang, Sampreetha Thampy, Wiley A. Dunlap-Shohl, Weijie Xu, Yangzi Zheng, Fong-Yi Cao, Yen-Ju Cheng, Anton V. Malko, David B. Mitzi and Julia W. P. Hsu
Mg Doped $CuCrO_2$ as Efficient Hole Transport Layers for Organic and Perovskite Solar Cells
Reprinted from: *Nanomaterials* **2019**, *9*, 1311, doi:10.3390/nano9091311 73

Han Wu, Zhong Ma, Zixia Lin, Haizeng Song, Shancheng Yan and Yi Shi
High-Sensitive Ammonia Sensors Based on Tin Monoxide Nanoshells
Reprinted from: *Nanomaterials* **2019**, *9*, 388, doi:10.3390/nano9030388 95

Dinesh Bhalothia, Yu-Jui Fan, Yen-Chun Lai, Ya-Tang Yang, Yaw-Wen Yang, Chih-Hao Lee and Tsan-Yao Chen
Conformational Effects of Pt-Shells on Nanostructures and Corresponding Oxygen Reduction Reaction Activity of Au-Cluster-Decorated NiO_x
Reprinted from: *Nanomaterials* **2019**, *9*, 1003, doi:10.3390/nano9071003 105

Chao-Feng Liu, Xin-Gui Tang, Lun-Quan Wang, Hui Tang, Yan-Ping Jiang, Qiu-Xiang Liu, Wen-Hua Li and Zhen-Hua Tang
Resistive Switching Characteristics of HfO_2 Thin Films on Mica Substrates Prepared by Sol-Gel Process
Reprinted from: *Nanomaterials* **2019**, *9*, 1124, doi:10.3390/nano9081124 121

Marco Moreira, Emanuel Carlos, Carlos Dias, Jonas Deuermeier, Maria Pereira, Pedro Barquinha, Rita Branquinho, Rodrigo Martins and Elvira Fortunato
Tailoring IGZO Composition for Enhanced Fully Solution-Based Thin Film Transistors
Reprinted from: *Nanomaterials* **2019**, *9*, 1273, doi:10.3390/nano9091273 133

About the Special Issue Editor

Julia W. P. Hsu is Professor of Materials Science and Engineering in the Erik Jonsson School of Engineering and Computer Science of the University of Texas at Dallas (UTD) and holds the Texas Instruments Distinguished Chair in Nanoelectronics. She received her B. S. E. degree in Chemical Engineering from Princeton University in 1985 and her Ph.D. degree in Physics from Stanford University in 1991. After a two-year postdoc at Bell Labs, she joined the faculty at the University of Virginia (UVA) as an Assistant Professor of Physics, earning tenure there in 1997. In 1999, she returned to Bell Labs as a Member of Technical Staff. Prior to coming to UTD, she was a Principal Member of Technical Staff at Sandia National Laboratories in Albuquerque NM from 2003 to 2010.

Prof. Hsu's research is in the area of nanoscale materials physics and interfacial phenomena at the interfaces of dissimilar materials. She has done extensive work on the spatially resolved characterization of electronic and photonic materials and devices using scanning probe techniques. The material systems she has studied are wide ranging, including metals and alloys; group IV, III–V, and II–VI semiconductors; polymers; nanocomposites; and oxides. Her work focuses on how macroscopic materials properties or device characteristics are affected by local materials chemistry or materials processing. Her recent research focuses on nanomaterials for optoelectronic and energy applications, including organic photovoltaics, nanomaterial synthesis, solution processing of inorganic nanocrystals and thin films, the synthesis and processing of few-layer transition metal dichalcogenides, electrical and optoelectronic studies of solar cells and transistors, earth-abundant oxides for clean air treatment, and low-temperature high-throughput processing of flexible electronics. Prof. Hsu has published over 200 journal papers, has been granted five patents, and has given over 180 invited talks.

Prof. Hsu is a Fellow of the American Physical Society (APS), the American Association for the Advancement of Science, and the Materials Research Society (MRS). She is a winner of a Hertz Foundation Fellowship, APS Apker Award, a National Science Foundation Young Investigator Award, and a Sloan Foundation Research Fellowship. Prof. Hsu currently serves on Department of Energy Basic Energy Science Advisory Committee. She was an August-Wilhelm Scheer Visiting Professor at Technische Universität München in 2018 and was recently awarded a Visiting Research Professorship at the University of Hong Kong. She was an organizer of the TMS Electronic Materials Conference and a co-chair for the Fall 2004 MRS meeting. She served as a Member-at-Large on the APS Division of Materials Physics Executive Committee (2004–2007), on the MRS Board of Directors (2005–2007), as the Treasurer and Chair of Operation Oversight Committee for the MRS (2006–2007), as chair of the MRS International Relations Committee from 2010 to 2011, and is a long-time member of MRS Meeting Assessment Subcommittee. She has held several key committee positions at UVA and UTD, and been on numerous review panels for funding agencies and on external advisory committees for research centers.

Editorial

Solution Synthesis, Processing, and Applications of Semiconducting Nanomaterials

Julia W. P. Hsu

Department of Materials Science and Engineering, University of Texas at Dallas, Richardson, TX 75080, USA; jwhsu@utdallas.edu

Received: 24 September 2019; Accepted: 4 October 2019; Published: 11 October 2019

Nanomaterials have contributed to the forefront of materials research in the past two decades, and are used today in sensors, solar cells, light emitting diodes, electronics, and biomedical devices. Solution synthesis and processing offer inexpensive, low-temperature, energy efficient, and environmentally friendly approaches that are desired especially for mass production or integration with plastic substrates. While metal nanoparticles, in particular gold, have been researched widely, semiconducting nanomaterials offer much greater versatility, functionality, and applications. The frontiers in synthesis include new compounds, reducing the size of nanomaterials, introducing new morphologies from assembly or templating, alternative green synthesis methods to reduce waste and energy, and surface functionalization. Furthermore, great challenges are encountered in processing nanomaterials from suspensions to thin films on different substrates for device applications. While many publications focus on synthesis and applications of solution-based nanomaterials, issues related to processing, e.g., solvent choice, surface compositions and ligands, particle–particle interaction, and deposition methods, are infrequently addressed. This Special Issue includes nine articles and one review, covering synthesis [1], novel processing [2–5], and a wide range of applications, such as solar cells [1,3,6], sensors [7], catalysis [8], and electronics [9,10].

Cu_2ZnSnS_4 (CZTS) is a kesterite material for solar cells based on earth abundant elements. An ecofriendly method to fabricate CZTS films is much sought after. Zhang et al. [1] adopted a wet ball milling method, which uses only nontoxic solvents. The films made from as-fabricated CZTS nanocrystal inks were followed by a rapid high-pressure sulfur annealing step to promote grain growth and crystallinity. They achieved solar cell efficiency of 6.2% (open circuit voltage: Voc = 633.3 mV, short circuit current: Jsc = 17.6 mA/cm^2, and fill factor: FF = 55.8%) with an area of 0.2 cm^2. Replacing sulfur with selenium in a kesterite material lowers its bandgap to better match the solar spectrum. Jiang et al. [3] investigated selenization treatment on $Cu_2Mg_{0.2}Zn_{0.8}Sn(S,Se)_4$ (CMZTSSe) films made by a sol–gel method. By controlling the selenization temperature and time, the crystallinity, film morphology, band gap, and hole concentration can be tuned. An optimized processing condition was reported.

Delafossite materials, with a chemical formula of AMO_2 (A = Cu or Ag and M is a trivalent ion), are rare p-type oxides. It was first reported in 1997 that $CuAlO_2$ was a true p-type transparent conducting oxide, which stimulated many activities in delafossite materials [11]. This Special Issue includes two articles on $CuCrO_2$ (CCO). Wu et al. [5] made novel Al_2O_3-CCO core-shell nanofibers and CCO hollow nanotubes. The CCO was grown on the surface of electrospun Al_2O_3 fibers by a solution method followed by annealing at 600 °C in a vacuum, resulting in Al_2O_3-CCO core-shell nanofibers. By applying sulfuric acid to the core-shell nanofibers, the authors showed that the Al_2O_3 cores were selectively removed while the rest of the original structures were preserved, hence producing CCO hollow nanotubes with an inner diameter of 70 nm and a wall thickness of 30 nm. Zhang et al. [6] introduced Mg as a dopant in CCO nanoparticles and examined the effects on the performance of solar cells using these materials as the hole transport layer (HTL). CCO nanoparticles have been shown as an efficient HTL in dye sensitized solar cells [12], organic solar cells [13], and, recently, halide perovskite solar cells [14–17]. This work was the first demonstration of how CCO and Mg-doped CCO perform as efficient HTLs for a non-fullerene acceptor bulk heterojunction (BHJ) system. They also observed Mg doping results in a small but definitive increase in the short circuit current density for all active layer systems, including three different BHJs and halide perovskite.

In addition to CCO hollow nanotubes, complex nanostructures have many interesting properties and applications. Bhalothia et al. [8] reported on the synthesis and oxygen reduction reaction activity of Au-cluster-decorated NiO_x@Pt nanostructures. They found an impressive performance of these nanocatalysts compared to commercial benchmarks: A 17-times larger kinetic current and a 53-fold increase in specific activity. Wu et al. [7] fabricated ammonia sensors based on self-assembly SnO nanoshells via a solution method that can detect ammonia below 20 ppm with high selectivity. Last but not least, Yun and Paik [4] contributed a review to the Special Issue on self-assembly of inorganic nanocrystals into superlattice thin films and multiscale nanostructures. The paper includes diverse examples of highly ordered superlattices.

Intense pulsed light (IPL) irradiation shows promise in rapid material processing that is compatible with roll-to-roll manufacturing [18]. Nakamura et al. [2] synthesized Cu nitride nanoparticles as an ink and converted them to Cu wires using IPL processing. This approach allows the production of large quantities of printed circuit boards with less waste. IPL also has the potential for device fabrication on low-temperature plastic substrates. Other electronic applications in this Special Issue includes flexible HfO_2 memory devices and indium-gallium-zinc oxide (IGZO) thin film transistors (TFTs). Liu et al. [9] prepared a novel flexible $Au/HfO_2/Pt$ resistive random access memory devices on a mica substrate using a sol–gel process. Moreira et al. [10] investigated solution processed IGZO films to replace vacuum deposition to implement low-cost, high-performance electronic devices on flexible transparent substrates. They evaluated the influence of composition, thickness, and aging on the electrical properties of IGZO TFTs, using a solution combustion synthesis method with urea as fuel. The optimized TFT built on a solution-processed AlO_x dielectric showed a saturation mobility of 3.2 cm^2 V^{-1} s^{-1}, an on–off ratio of 10^6, a sub-threshold swing of 73 mV dec^{-1}, and a threshold voltage of 0.18 V, thus demonstrating promising features for low-cost circuit applications.

I hope *Nanomaterials* readers find these articles informative and interesting.

Funding: Texas Instruments Distinguished Chair in Nanoelectronics partially support JWPH's efforts in putting together this Special Issue.

Acknowledgments: The Guest Editor would like to thank all authors for submitting their work to the Special Issue and for its successful completion. A special recognition also goes to all the reviewers for their prompt responses and for making constructive suggestions that enhance the publication quality and impact. I am also grateful to Sandra Ma and the editorial assistants who made the Special Issue creation a smooth and efficient process.

Conflicts of Interest: The authors declare no conflict of interest.

References

1. Zhang, X.; Fu, E.; Wang, Y.; Zhang, C. Fabrication of Cu_2ZnSnS_4 (CZTS) Nanoparticle Inks for Growth of CZTS Films for Solar Cells. *Nanomaterials* **2019**, *9*, 336. [CrossRef] [PubMed]
2. Nakamura, T.; Cheong, H.J.; Takamura, M.; Yoshida, M.; Uemura, S. Suitability of Copper Nitride as a Wiring Ink Sintered by Low-Energy Intense Pulsed Light Irradiation. *Nanomaterials* **2018**, *8*, 617. [CrossRef] [PubMed]
3. Jiang, D.; Zhang, Y.; Sui, Y.; He, W.; Wang, Z.; Yang, L.; Wang, F.; Yao, B. Investigation on the Selenization Treatment of Kesterite $Cu_2Mg_{0.2}Zn_{0.8}Sn(S,Se)_4$ Films for Solar Cell. *Nanomaterials* **2019**, *9*, 946. [CrossRef] [PubMed]
4. Yun, H.; Paik, T. Colloidal Self-Assembly of Inorganic Nanocrystals into Superlattice Thin-Films and Multiscale Nanostructures. *Nanomaterials* **2019**, *9*, 1243. [CrossRef] [PubMed]
5. Wu, H.-J.; Fan, Y.-J.; Wang, S.-S.; Sakthinathan, S.; Chiu, T.-W.; Li, S.-S.; Park, J.-H. Preparation of $CuCrO_2$ Hollow Nanotubes from an Electrospun Al_2O_3 Template. *Nanomaterials* **2019**, *9*, 1252. [CrossRef] [PubMed]

6. Zhang, B.; Thampy, S.; Dunlap-Shohl, W.A.; Xu, W.; Zheng, Y.; Cao, F.-Y.; Cheng, Y.-J.; Malko, A.V.; Mitzi, D.B.; Hsu, J.W.P. Mg Doped CuCrO$_2$ as Efficient Hole Transport Layers for Organic and Perovskite Solar Cells. *Nanomaterials* **2019**, *9*, 1311. [CrossRef] [PubMed]
7. Wu, H.; Ma, Z.; Lin, Z.; Song, H.; Yan, S.; Shi, Y. High-Sensitive Ammonia Sensors Based on Tin Monoxide Nanoshells. *Nanomaterials* **2019**, *9*, 388. [CrossRef] [PubMed]
8. Bhalothia, D.; Fan, Y.J.; Lai, Y.C.; Yang, Y.T.; Yang, Y.W.; Lee, C.H.; Chen, T.Y. Conformational Effects of Pt-Shells on Nanostructures and Corresponding Oxygen Reduction Reaction Activity of Au-Cluster-Decorated NiO$_x$@Pt Nanocatalysts. *Nanomaterials* **2019**, *9*, 1003. [CrossRef] [PubMed]
9. Liu, C.-F.; Tang, X.-G.; Wang, L.-Q.; Tang, H.; Jiang, Y.-P.; Liu, Q.-X.; Li, W.-H.; Tang, Z.-H. Resistive Switching Characteristics of HfO$_2$ Thin Films on Mica Substrates Prepared by Sol-Gel Process. *Nanomaterials* **2019**, *9*, 1124. [CrossRef] [PubMed]
10. Moreira, M.; Carlos, E.; Dias, C.; Deuermeier, J.; Pereira, M.; Barquinha, P.; Branquinho, R.; Martins, R.; Fortunato, E. Fortunato Tailoring IGZO Composition for Enhanced Fully Solution-Based Thin Film Transistors. *Nanomaterials* **2019**, *9*, 1273. [CrossRef] [PubMed]
11. Kawazoe, H.; Yasukawa, M.; Hyodo, H.; Kurita, M.; Yanagi, H.; Hosono, H. P-type electrical conduction in transparent thin films of CuAlO$_2$. *Nature* **1997**, *389*, 939–942. [CrossRef]
12. Xiong, D.; Xu, Z.; Zeng, X.; Zhang, W.; Chen, W.; Xu, X.; Wang, M.; Cheng, Y.-B. Hydrothermal synthesis of ultrasmall CuCrO$_2$ nanocrystal alternatives to NiO nanoparticles in efficient p-type dye-sensitized solar cells. *J. Mater. Chem.* **2012**, *22*, 24760–24768. [CrossRef]
13. Wang, J.; Lee, Y.-J.; Hsu, J.W.P. Sub-10 nm copper chromium oxide nanocrystals as a solution processed p-type hole transport layer for organic photovoltaics. *J. Mater. Chem. C* **2016**, *4*, 3607–3613. [CrossRef]
14. Zhang, H.; Wang, H.; Zhu, H.; Chueh, C.-C.; Chen, W.; Yang, S.; Jen, A.K.Y. Low-Temperature Solution-Processed CuCrO$_2$ Hole-Transporting Layer for Efficient and Photostable Perovskite Solar Cells. *Adv. Energy Mater.* **2018**, *8*, 1702762. [CrossRef]
15. Dunlap-Shohl, W.A.; Daunis, T.B.; Wang, X.; Wang, J.; Zhang, B.; Barrera, D.; Yan, Y.; Hsu, J.W.P.; Mitzi, D.B. Room-temperature fabrication of a delafossite CuCrO$_2$ hole transport layer for perovskite solar cells. *J. Mater. Chem. A* **2018**, *6*, 469–477. [CrossRef]
16. Yang, B.; Ouyang, D.; Huang, Z.; Ren, X.; Zhang, H.; Choy, W.C.H. Multifunctional Synthesis Approach of In:CuCrO$_2$ Nanoparticles for Hole Transport Layer in High-Performance Perovskite Solar Cells. *Adv. Funct. Mater.* **2019**, *29*, 1902600. [CrossRef]
17. Akin, S.; Liu, Y.; Dar, M.I.; Zakeeruddin, S.M.; Grätzel, M.; Turan, S.; Sonmezoglu, S. Hydrothermally Processed CuCrO$_2$ Nanoparticles as an Inorganic Hole Transporting Material for Low-cost Perovskite Solar Cells with Superior Stability. *J. Mater. Chem. A* **2018**, *6*, 20327–20337. [CrossRef]
18. Schroder, K.A.; McCool, S.C.; Furlan, W.F. Broadcast Photonic Curing of Metallic Nanoparticle Films. *NSTI-Nanotech May* **2006**, *3*, 198–201.

© 2019 by the author. Licensee MDPI, Basel, Switzerland. This article is an open access article distributed under the terms and conditions of the Creative Commons Attribution (CC BY) license (http://creativecommons.org/licenses/by/4.0/).

Article

Fabrication of Cu_2ZnSnS_4 (CZTS) Nanoparticle Inks for Growth of CZTS Films for Solar Cells

Xianfeng Zhang [1,*], Engang Fu [2], Yuehui Wang [1] and Cheng Zhang [1]

1. Zhongshan Institute, University of Electronic Science and Technology of China, Zhongshan 528402, Guangdong, China; wangzsedu@126.com (Y.W.); aqian2006@gmail.com (C.Z.)
2. State Key Laboratory of Nuclear Physics and Technology, School of Physics, Peking University, Beijing 100871, China; efu@pku.edu.cn
* Correspondence: zhangxf07@gmail.com; Tel.: +86-760-8831-3456

Received: 8 January 2019; Accepted: 24 February 2019; Published: 2 March 2019

Abstract: Cu_2ZnSnS_4 (CZTS) is a promising candidate material for photovoltaic applications; hence, ecofriendly methods are required to fabricate CZTS films. In this work, we fabricated CZTS nanocrystal inks by a wet ball milling method, with the use of only nontoxic solvents, followed by filtration. We performed centrifugation to screen the as-milled CZTS and obtain nanocrystals. The distribution of CZTS nanoparticles during centrifugation was examined and nanocrystal inks were obtained after the final centrifugal treatment. The as-fabricated CZTS nanocrystal inks were used to deposit CZTS precursors with precisely controlled CZTS films by a spin-coating method followed by a rapid high pressure sulfur annealing method. Both the grain growth and crystallinity of the CZTS films were promoted and the composition was adjusted from S poor to S-rich by the annealing. XRD and Raman characterization showed no secondary phases in the annealed film, the absence of the detrimental phases. A solar cell efficiency of 6.2% (open circuit voltage: V_{oc} = 633.3 mV, short circuit current: J_{sc} = 17.6 mA/cm^2, and fill factor: FF = 55.8%) with an area of 0.2 cm^2 was achieved based on the annealed CZTS film as the absorber layer.

Keywords: Cu_2ZnSnS_4 solar cell; ball milling; nano-ink; annealing

---

1. Introduction

In recent years, kesterite Cu_2ZnSnS_4 (CZTS) and $Cu_2ZnSn(S, Se)_4$ (CZTSSe) solar cells have drawn attention because of their promise as an absorbing layer for applications in thin-film photovoltaics owing to its low cost, nontoxicity and earth abundance of its elemental components as well as an adjustable bandgap [1–3]. One advantage of CZTS over other kinds of chalcopyrite-related solar cells is its suitability for achieving high efficiency solar cells through nonvacuum fabrication methods. Furthermore, the world record conversion efficiency of CZTSSe solar cells is currently 12.6% [4] based on a hydrazine pure solution approach. There have been several reports on the fabrication of CZTS or CZTSSe solar cells. Both vacuum methods, such as sputtering [5], coevaporation [6], epitaxial methods [7], and nonvacuum methods [8–11], have been reported. Nonvacuum methods are lower in cost and more suitable for mass production than are vacuum methods. Among nonvacuum methods, the highest conversion efficiency CZTS solar cells are based on molecular precursor solutions or nanoparticle dispersions [12,13]. Although these kinds of fabrication methods are appealing because of their low complexity, low-cost, and scalability, such methods are complicated by the need for toxic solvents or metal–organic solutions that contain large amounts of organic contaminants, which induce cracking during the following annealing process [14,15]. The use of the toxic and unstable solvent hydrazine requires all processes for ink and film preparation to be performed under an inert atmosphere. As a result, it is difficult to adapt this approach to low-cost and large-scale solar cell fabrication.

In this work, we report a simple technique for fabricating CZTS nanoparticles ink by a wet ball milling method using nontoxic ethanol and 2-(2-ethoxyethoxy) ethanol as the solvents. A similar study has been reported by Woo, et al.; an efficiency of 7% was achieved [16]. However, the use of CZTS powder to fabricate CZTS film is expected to have the following benefits. (1) The fabrication process is simpler. (2) The growth of grains will be promoted because the grain boundary meltdown temperature is lowered. (3) Stoichiometric film compositions are easier to obtain because the chemical reactions are less complicated. The use of nontoxic solvents is more cost-effective and environment friendly, which is important for practical photovoltaic applications. The ink was used to fabricate CZTS thin films (precursors) by a spin-coating method, followed by annealing the precursor in a sulfur-rich atmosphere. The commercial CZTS powder was obtained from Mitsui Kinzoku, and detailed information of its characteristics is currently unavailable. The sulfur vapor not only prevents the formation of volatile Sn–S compounds but also supplies S atoms to make the CZTS films sulfur-rich, which is a requirement for high performance solar cells. The procedure for fabricating CZTS films from CZTS powder is reported in detail in this paper.

2. Experiment Details

2.1. Sample Preparation

Figure 1a–c illustrates the process for fabricating CZTS nanoparticle ink. In the ball milling system, a 1-mm ball, 50-μm ceramic balls, and CZTS powder were mixed together in the mill pot. A 5-mL portion of ethanol was added to improve the wet milling effect. Figure 1a shows a schematic diagram of the milling system. The milling pots were rotated along their own axis together with the base plate. The milling process was performed for 40 h. After ball milling, the whole mixture was strained through a filter screen to obtain particles smaller than 32 μm, and nontoxic ethanol and 2-(2-ethoxyethoxy) ethanol were used to wash the milling ball to increase nanoparticle recovery, as shown in Figure 1b. Through this procedure, the milling balls and large particles of CZTS (>32 μm) were removed whereas a mixture of relatively small CZTS particles (<32 μm) and the solvents were retained. We used 2-(2-ethoxyethoxy) ethanol as a dispersion agent to prevent coagulation of the nanoparticles, and ethanol was used to reduce the viscosity of the solvent and promote precipitation of large particles during the following centrifugation. The resulting solution was then ultrasonically processed to disperse the particles in the solvents for 1 h.

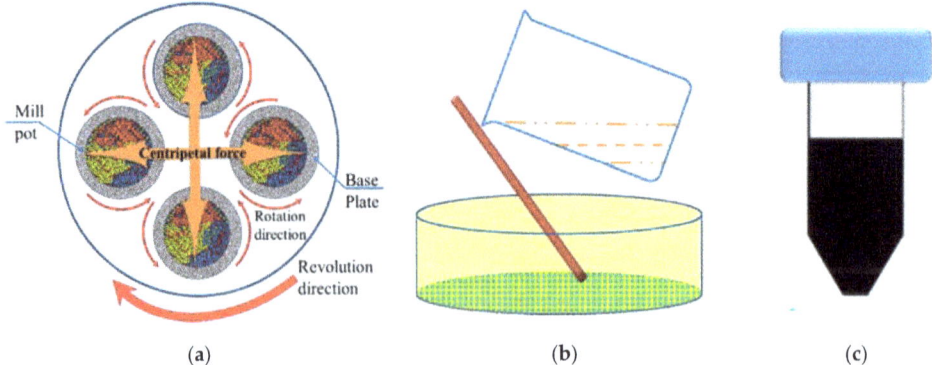

Figure 1. Fabrication process of Cu_2ZnSnS_4 (CZTS) nanoparticle ink: (a) schematic of ball milling machine, (b) filtration of CZTS particles smaller than 32 μm, (c) nanoparticle ink of CZTS.

The as-prepared mixture of CZTS and solvents was first centrifuged at a low speed 1500 rpm to remove particles over several μm in size. The precipitate was disposed of and the upper layer of the solution was decanted for further centrifugal treatment. The aforementioned processes were repeated three times at a higher speed of 6000 rpm and the nanoparticles were obtained. The nanoparticle ink was obtained with a concentration of 200 mg/mL by adjusting the quantity of ethanol. The nanoparticle ink was then used to fabricate CZTS precursors by spin-coating. Figure 2a shows a schematic diagram of the spin-coating system. The substrate was rotated at a speed of 2000 rpm and the CZTS ink was dripped on at a speed of 5 µL/min. The final CZTS precursor film showed a thickness of 1–1.5 µm. Finally, the precursors were annealed in a sulfur-rich atmosphere to improve the grain size and crystallinity. The sulfurization process was conducted by sealing the precursor and powdered sulfur into a vacuum quartz tube with a length of 15 cm, which was placed in the annealing furnace (FP410, Yamato Company, Tokyo, Japan), as shown in Figure 1b. The furnace was heated to 600 °C within 15 min and the vapor pressure of sulfur was approximately 0.1 atm. The annealing process was performed for 20 min after the system achieved 600 °C. Then the sample was allowed to cool to room temperature naturally.

Figure 2. Fabrication process of CZTS films: (**a**) fabrication of CZTS precursor by spin-coating and (**b**) the sulfur vapor annealing process.

A typical structure of a CZTS solar cell is shown in Figure 3. The as-grown CZTS film was used as the absorbing layer. A CdS layer with a thickness of 50 nm was fabricated by a chemical bath deposition method as the buffer layer. Intrinsic ZnO with a thickness of 100 nm and B-doped ZnO with a thickness of 400 nm were then sputtered as the window layer. To measure the performance of the solar cell, an Al grid was evaporated as the front electrode.

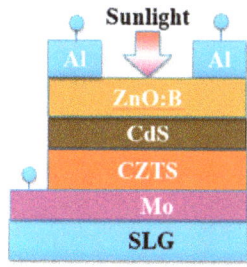

Figure 3. Structure of CZTS solar cell.

2.2. Characterization

The morphology of the annealed CZTS films was characterized with a scanning electron microscope (SEM, JSM-7001F, Tokyo, Japan) equipped with a JED-2300T energy dispersive spectroscopy (EDS) system (Tokyo, Japan) operating at an acceleration voltage of 10 kV. EDS, for compositional analysis, was measured at an acceleration voltage of 15 kV. The grain size distribution was measured with a transmission electron microscope (TEM, JEOL JEM-2100F, Tokyo, Japan). X-ray diffraction (XRD) analysis was performed with a Rigaku SmartLab2 with a Cu-K source and the generator was set to 20 mA and 40 kV. Raman measurements were performed with a RENISHAW-produced inVia RefleX type Raman spectrometer equipped with an Olympus microscope with a 1000 magnification lens at room temperature. The excitation laser line was 532 nm. The solar cell performance was measured with a 913 CV type current–voltage (J–V) tester (AM1.5) provided by a EKO (LP-50B, Tokyo, Japan) solar simulator. The simulator was calibrated with a standard GaAs solar cell to obtain the standard illumination density (100 mW/cm^2).

3. Results and Discussion

3.1. Centrifugation to Obtain CZTS Nanoparticle Ink

Figure 4a–e shows TEM images of the CZTS particle distribution of the dispersion subjected to different centrifugation conditions. Figure 4a shows the distribution of CZTS particles for the CZTS dispersion without a centrifugal treatment. The small particles and large particles agglomerated together to form large clusters such that the boundaries between particles became unclear and it was not possible to tell the size of the particles; hence, the larger and smaller particles and nanoparticles were not separated. Figure 4b shows an TEM image of the CZTS ink centrifuged for 10 min at 1500 rpm. A portion of the large particles was removed, which reduced the agglomeration. The particle boundaries were clear; however, particles larger than several hundred nm remained. To further reduce the size of the particles, the dispersion was centrifuged at a high speed of 6000 rpm for 10, 20, and 30 min. The results are shown in Figure 4c–e, respectively. The sample shown in Figure 4c, had the largest particles (in the range of 100 to 200 nm) and almost no agglomeration was observed. In sample (d), particles remaining in the dispersion were smaller than 100 nm, indicating that nanoparticles were obtained. The particle size of sample (e) was in the range of 50 to 100 nm, which indicated that after the treatment to obtain sample (d), the particle size of the dispersion was no longer affected by centrifugation because of the limitations of final particle sizes generated by ball milling processes.

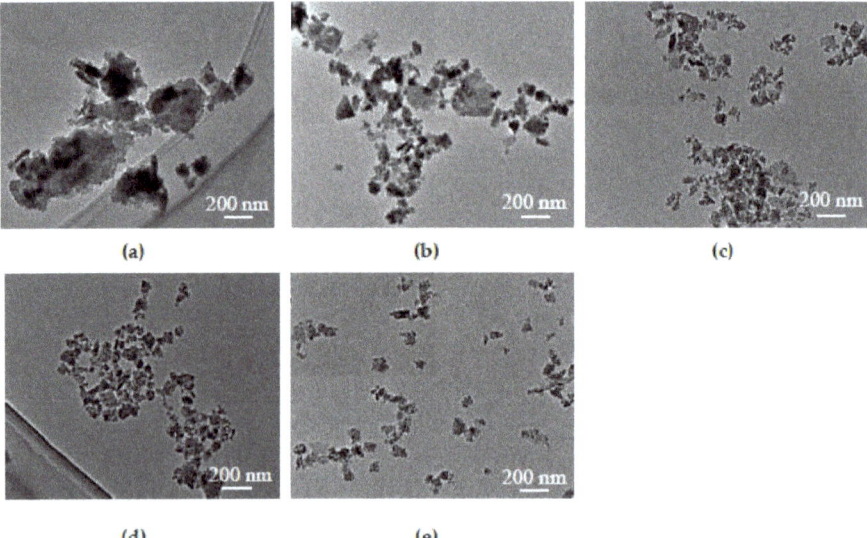

Figure 4. TEM images of CZTS dispersions with different centrifugal conditions. Distribution of CZTS particle size in inks with different centrifugal conditions (**a**) Without centrifugation; (**b**) 1500 rpm for 10 min; (**c**) 6000 rpm for 10 min; (**d**) 6000 rpm for 20 min; and (**e**) 6000 rpm for 30 min.

3.2. Deposition of CZTS Precursors

The CZTS nanoparticle inks were used to deposit the CZTS precursors on glass substrates by a spin-coating method. The speed of the substrate was approximately 2000 rpm and 5 µL of CZTS ink was dripped at the center of the substrate for each drop, which was repeated 10 times to obtain a film with a thickness of 1–1.5 μm. Figure 5a–c shows the surface morphology of the CZTS film with different magnifications. The SEM image showed a compact morphology with grains smaller than 100 nm without cracks and no large particles were observed. The specific grain size could not be measured because of the small boundaries between grains. Because the precursor was only grown at room temperature, an additional high-temperature treatment was necessary to improve the grain size and crystallinity of the film.

Figure 5. SEM images of a CZTS film deposited from the as-fabricated CZTS nanoparticle inks under different magnifications: (**a**) 20000×; (**b**) 10000×; and (**c**) 5000×.

3.3. Annealing of the Precursor

To induce grain growth and reduce the residual organic impurities, the CZTS precursor was annealed in an atmosphere with a high sulfur vapor pressure for 20 min at a temperature of 600 °C. Figure 6a,b shows the surface and cross-sectional SEM images of the CZTS films after annealing, respectively. Comparing the precursor morphology, as shown in Figure 5, the grain size increased markedly. The final grain size ranged from several hundred nm to several μm and cracks begin to appear between the grains, either because of grain growth or decomposition of the CZTS particles. According to the cross-sectional image (Figure 6b), the grains extended throughout the film in the thickness direction, which is expected for high-quality films. However, cracks stretching from the surface to the bottom of the film were also observed (marked by the red arrow), indicating the low density of the film. One explanation for this cracking was reported by Scragg, et al. owing to decomposition of CZTS film, as shown in following reactions (1) and (2) [17].

$$Cu_2ZnSnS_4 \rightleftharpoons Cu_2S(s) + ZnS(s) + SnS(s) + 1/2S_2(g) \tag{1}$$

$$SnS(s) \rightleftharpoons SnS(g) \tag{2}$$

One solution to overcome this issue is to reduce the annealing temperature to prevent equilibrium (1) from shifting to the right and extending the annealing time to ensure maintain the crystallinity.

 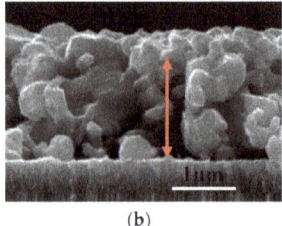

(a) (b)

Figure 6. (a) Surface and (b) cross-section of an annealed CZTS film annealed at 600 °C in a S-rich atmosphere.

To make a comparison, CZTS film using centrifugation condition: 1500 rmp for 10 min was also annealed with the same annealing condition and completed solar cell structure (Please refer to the Supplementary Materials).

Table 1 shows the composition of the CZTS precursor and annealed film, as determined by energy-dispersive X-ray spectroscopy (EDX). The precursor had a sulfur composition less than 50% whereas the sulfur content increased to 50.5% after annealing, indicating that the film was converted from sulfur poor to sulfur-rich, which produces p-type CZTS films. It has been widely reported that Zn-rich (Zn/Sn > 1.0) films are required for fabricating high-performance CZTS solar cells [18,19], meaning that the composition of our CZTS films needed to be adjusted. One possible way to adjust the film to Zn-rich is to fabricate a thin layer of ZnS nanoparticles between the CZTS precursor and Mo back-contact, such that in the following annealing step, both Zn and S will be supplemented.

Table 1. Composition of precursor and annealed film as measured by energy-dispersive X-ray spectroscopy (EDX).

	Cu (%)	Zn (%)	Sn (%)	S (%)	Zn/Sn	Cu/(Zn + Sn)
Precursor	25.4	9.9	15.6	49.1	0.63	1.00
Annealed CZTS film	24.7	9.8	15.1	50.5	0.65	0.99

Figure 7 shows the XRD patterns of the precursor and annealed film of CZTS. The crystallinity was also improved by high-temperature annealing. The sulfurization process induced sharpening and strengthening of the peaks. All the peaks of the precursor and the annealed film were assigned to kesterite CZTS. No peaks of secondary phases, such as ZnS and Cu_2S, which easily form at high temperatures [20], were detected by XRD. However, XRD alone is incapable of identifying small amounts of secondary phases because of its detection limits. To complement this method, we also performed Raman measurements to confirm the absence of secondary phases. Raman spectra of the precursor and annealed CZTS thin films are shown in Figure 8. The lower spectrum shows the annealed CZTS film with peak fitting by a Lorentzian curve. According to the figure, the precursor showed one peak at 330 cm^{-1}, corresponding to the A mode of kesterite CZTS. The annealed film exhibited a typical Raman spectrum of kesterite CZTS films with three peaks at 285, 330, and 369 cm^{-1}, corresponding to the two A symmetry modes and a B symmetry mode of the CZTS kesterite structure, respectively [21,22]. This result also indicated that no secondary phases are observed after the annealing process.

The annealed CZTS films were used to fabricate complete solar cell structures. Solar cell performance was evaluated under standard conditions. The conversion efficiency of three cells on the same sample was measured as shown in Table 2. The solar cell ranged from 2.5% to 6.2%, indicating ununiform solar cell performance due to the poor film quality as shown in Figure 7.

Table 2. Performance of CZTS solar cells.

Sample No.	E_{ff}	V_{oc} (mV)	J_{sc} (mA/cm^2)	FF (%)
1	6.2	633.3	17.6	55.8
2	4.3	578.2	15.3	48.6
3	2.5	497.1	12.2	41.2

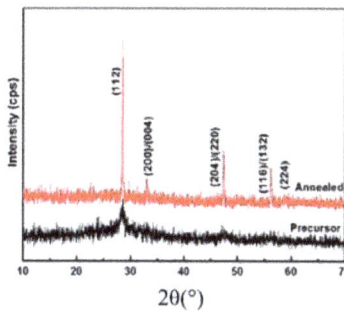

Figure 7. XRD of precursor and annealed CZTS film.

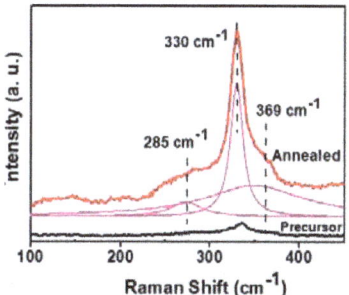

Figure 8. Raman spectrums of precursor and annealed CZTS film with fitting of the peaks using Lorentzian curve.

Figure 9 shows dark and light I–V curves of solar cell using annealed CZTS film as the absorber layer with best solar cell performance. The photovoltaic device exhibited an efficiency of 6.2%, with V_{oc} = 633.3 mV, J_{sc} = 17.6 mA/cm^2, and FF = 55.8%, for an area of 0.20 cm^2.

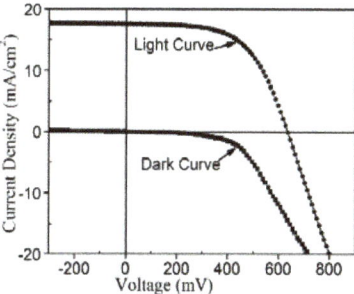

Figure 9. J–V curve of CZTS.

Figure 10 shows the external quantum efficiency (EQE) curve of the CZTS solar cell. Over the visible range of the solar spectrum, the maximum QE was less than 60%, indicating strong recombination. The QE curve decreased sharply in the infrared region at 770 nm, which is the CZTS absorption edge. Thus, the calculated bandgap of the CZTS films was approximately 1.61 eV. The features near 510 nm and 380 nm correspond to the absorption edges of the CdS and ZnO layers [23,24], which are commonly used CdS buffer and ZnO window layers.

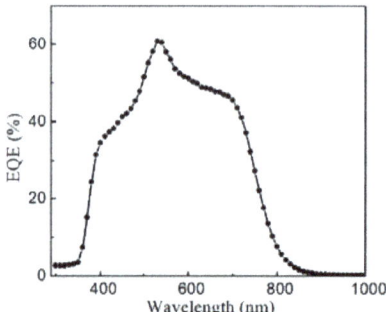

Figure 10. External quantum efficiency (EQE) of CZTS solar cell.

On the basis of the EQE data of a solar cell, J_{sc} was calculated as [25]

$$J_{sc} = q \int_0^\infty QE(E) b_s(E, T_s) dE \qquad (3)$$

where, q is the elementary charge, QE is the quantum efficiency, and b_s is solar flux or irradiation. For an air mass of 1.5, the data is available from Ref. [26]. On the basis of Equation (3), Figure 10, and the solar irradiation spectrum, J_{sc} of the CZTS solar cells was calculated to be 14.2 mA/cm², because the J-V curve represents the real performance of a photovoltaic device. The slight deviation of J_{sc} calculated from the QE curve can be explained by the fact that the QE measurement is performed at a single wavelength with a much lower intensity than one-sun irradiation.

4. Conclusions

We synthesized a CZTS nanoparticle ink by a wet ball milling method together with centrifugation treatments based on only nontoxic solvents. The ink was then used to deposit CZTS precursor films by a spin-coating method, which led to extremely flat surfaces with high-uniformity. The precursor was annealed at a high temperature of 600 °C under a sulfur atmosphere and the grain size increased to approximately 1 μm from the original size of less than 100 nm. Both the composition and crystallinity of the CZTS film were markedly improved by annealing. The absence of secondary phase formation during the annealing process was confirmed by XRD and Raman analysis. A solar cell efficiency of 6.2% (V_{oc} = 633.3 mV, J_{sc} = 17.6 mA/cm², and FF = 55.8%) with an area of 0.2 cm² was achieved using annealed CZTS film as the light absorbing layer. To improve solar cell performance, it is necessary to increase grain size, improve crystallinity, and reduce defects in the film. Because the fabrication process of CZTS features a complex growth mechanism, the formation of secondary phases should be checked to confirm film quality, which directly affects solar cell performance.

Supplementary Materials: The following are available online at http://www.mdpi.com/2079-4991/9/3/336/s1, Figure S1: (a) Surface and (b) cross-section of an annealed CZTS film using centrifugation condition: 6000 rpm for 20 min. (c) Surface Morphology of an annealed CZTS film fabricated with CZTS ink using centrifugation condition: 1500 rpm for 10 min. The annealing was conducted at a temperature of 600 °C in S rich atmosphere, Figure S2: J-V curve of CZTS solar cells for (a) centrifugation 6000 rpm for 20 min; (b) 1500 rpm for 10 min, Table S1: Solar cell performance of CZTS solar cells.

Author Contributions: Conceptualization, X.Z., part of the characterization: E.F.; funding acquisition, Y.W.; draft review and editing, C.Z.

Acknowledgments: Part of the work was financially supported by Grant for special Research Project of Zhongshan Institute (funding No. 417YKQ10). This work was also financially supported by National Science Foundation of China under grants of (61302044, 61671140). We thank Edanz Group for editing a draft of this manuscript.

Conflicts of Interest: The authors declare no conflict of interest. The funders contribute in collecting results and editing the manuscript.

References

1. Roelofs, K.E.; Guo, Q.J.; Subramoney, S.; Caspar, J.V. Investigation of local compositional uniformity in $Cu_2ZnSn(S,Se)_4$ thin film solar cells prepared from nanoparticle inks. *J. Mater. Chem. A* **2014**, *2*, 13464–13470. [CrossRef]
2. Ravindiran, M.; Praveenkuma, C. Status review and the future prospects of CZTS based solar cell—A novel approach on the device structure and material modeling for CZTS based photovoltaic device. *Renew. Sustain. Energy Rev.* **2018**, *94*, 317–329. [CrossRef]
3. Fella, C.M.; Uhl, A.R.; Romanyuk, Y.E.; Tiwari, A.N. $Cu_2ZnSnSe_4$ absorbers processed from solution deposited metal salt precursors under different selenization conditions. *Phys. Status Solidi A* **2012**, *209*, 1043–1048. [CrossRef]
4. Wang, W.; Winkler, M.T.; Gunawan, O.; Gokmen, T.; Todorov, T.K.; Zhu, Y.; Mitzi, D.B. Device Characteristics of CZTSSe Thin-Film Solar Cells with 12.6% Efficiency. *Adv. Energy Mater.* **2014**, *4*, 1–5. [CrossRef]
5. Seol, J.S.; Lee, S.Y.; Lee, J.C.; Nam, H.D.; Kim, K.H. Electrical and optical properties of Cu_2ZnSnS_4 thin films prepared by rf magnetron sputtering process. *Sol. Energy Mater. Sol. C* **2003**, *75*, 155–162. [CrossRef]
6. Tanaka, T.; Kawasaki, D.; Nishio, M.; Gu, Q.X.; Ogawal, H. Fabrication of Cu_2ZnSnS_4 thin films by co-evaporation. *Phys. Status Solidi C* **2006**, *3*, 2844–2847. [CrossRef]
7. Song, N.; Young, M.; Liu, F.Y.; Erslev, P.; Wilson, S.; Harvey, S.P.; Teeter, G.; Huang, Y.D.; Hao, X.J.; Green, M.A. Epitaxial Cu_2ZnSnS_4 thin film on Si (111) 4 degrees substrate. *Appl. Phys. Lett.* **2015**, *106*, 1–9. [CrossRef]
8. Tanaka, K.; Oonuki, M.; Moritake, N.; Uchiki, H. Cu_2ZnSnS_4 thin film solar cells prepared by non-vacuum processing. *Sol. Energy Mater. Sol. C* **2009**, *93*, 583–587. [CrossRef]
9. Woo, K.; Kim, Y.; Moon, J. A non-toxic, solution-processed, earth abundant absorbing layer for thin-film solar cells. *Energy Environ. Sci.* **2012**, *5*, 5340–5345. [CrossRef]
10. Patel, M.; Mukhopadhyay, I.; Ray, A. Structural, optical and electrical properties of spray-deposited CZTS thin films under a non-equilibrium growth condition. *J. Phys. D Appl. Phys.* **2012**, *45*, 1–10. [CrossRef]
11. Cao, Y.Y.; Denny, M.S.; Caspar, J.V.; Farneth, W.E.; Guo, Q.J.; Ionkin, A.S.; Johnson, L.K.; Lu, M.J.; Malajovich, I.; Radu, D.; et al. High-Efficiency Solution-Processed $Cu_2ZnSn(S,Se)_4$ Thin-Film Solar Cells Prepared from Binary and Ternary Nanoparticles. *J. Am. Chem. Soc.* **2012**, *134*, 15644–15647. [CrossRef] [PubMed]
12. Shin, B.; Gunawan, O.; Zhu, Y.; Bojarczuk, N.A.; Chey, S.J.; Guha, S. Thin film solar cell with 8.4% power conversion efficiency using an earth-abundant Cu_2ZnSnS_4 absorber. *Prog. Photovolt.* **2013**, *21*, 72–76. [CrossRef]
13. Todorov, T.K.; Reuter, K.B.; Mitzi, D.B. High-Efficiency Solar Cell with Earth-Abundant Liquid-Processed Absorber. *Adv. Mater.* **2010**, *22*, E156–E159. [CrossRef] [PubMed]
14. Pawar, B.S.; Pawar, S.M.; Shin, S.W.; Choi, D.S.; Park, C.J.; Kolekar, S.S.; Kim, J.H. Effect of complexing agent on the properties of electrochemically deposited Cu_2ZnSnS_4 (CZTS) thin films. *Appl. Surf. Sci.* **2010**, *257*, 1786–1791. [CrossRef]
15. Akhavan, V.A.; Goodfellow, B.W.; Panthani, M.G.; Steinhagen, C.; Harvey, T.B.; Stolle, C.J.; Korgel, B.A. Colloidal CIGS and CZTS nanocrystals: A precursor route to printed photovoltaics. *J. Solid State Chem.* **2012**, *189*, 2–12. [CrossRef]
16. Woo, K.; Kim, Y.; Yang, W.; Kim, K.; Kim, I.; Oh, Y.; Kim, J.Y.; Moon, J. Band-gap-graded $Cu_2ZnSn(S_{1-x}Se_x)(4)$ Solar Cells Fabricated by an Ethanol-based, Particulate Precursor Ink Route. *Sci. Rep.* **2013**, *3*, 1–7. [CrossRef] [PubMed]
17. Scragg, J.J.; Ericson, T.; Kubart, T.; Edoff, M.; Platzer-Bjorkman, C. Chemical Insights into the Instability of Cu2ZnSnS4 Films during Annealing. *Chem. Mater.* **2011**, *23*, 4625–4633. [CrossRef]

18. Fairbrother, A.; Garcia-Hemme, E.; Izquierdo-Roca, V.; Fontane, X.; Pulgarin-Agudelo, F.A.; Vigil-Galan, O.; Perez-Rodriguez, A.; Saucedo, E. Development of a Selective Chemical Etch to Improve the Conversion Efficiency of Zn-Rich Cu_2ZnSnS_4 Solar Cells. *J. Am. Chem. Soc.* **2012**, *134*, 8018–8021. [CrossRef] [PubMed]
19. Ilari, G.M.; Fella, C.M.; Ziegler, C.; Uhl, A.R.; Romanyuk, Y.E.; Tiwari, A.N. Cu_2ZnSnS_4 solar cell absorbers spin-coated from amine-containing ether solutions. *Sol. Energy Mater. Sol. Cells* **2012**, *104*, 125–130. [CrossRef]
20. Redinger, A.; Berg, D.M.; Dale, P.J.; Siebentritt, S. The Consequences of Kesterite Equilibria for Efficient Solar Cells. *J. Am. Chem. Soc.* **2011**, *133*, 3320–3323. [CrossRef] [PubMed]
21. Khare, A.; Himmetoglu, B.; Johnson, M.; Norris, D.J.; Cococcioni, M.; Aydil, E.S. Calculation of the lattice dynamics and Raman spectra of copper zinc tin chalcogenides and comparison to experiments. *J. Appl. Phys.* **2012**, *111*, 083708. [CrossRef]
22. Fernandes, P.A.; Salome, P.M.P.; da Cunha, A.F.; Schubert, B.A. Cu_2ZnSnS_4 solar cells prepared with sulphurized dc-sputtered stacked metallic precursors. *Thin Solid Films* **2011**, *519*, 7382–7385. [CrossRef]
23. Wada, T.; Kohara, N.; Nishiwaki, S.; Negami, T. Characterization of the $Cu(In,Ga)Se_2$/Mo interface in CIGS solar cells. *Thin Solid Films* **2001**, *387*, 118–122. [CrossRef]
24. Ramanathan, K.; Contreras, M.A.; Perkins, C.L.; Asher, S.; Hasoon, F.S.; Keane, J.; Young, D.; Romero, M.; Metzger, W.; Noufi, R.; et al. Properties of 19.2% efficiency $ZnO/CdS/CuInGaSe_2$ thin-film solar cells. *Prog. Photovolt.* **2003**, *11*, 225–230. [CrossRef]
25. Nelson, J. *The Physics of Solar Cells*; Imperial College Press: London, UK; World Scientific Pub. Co.: River Edge, NJ, USA, 2003; p. 363.
26. Christians, J.A.; Manser, J.S.; Kamat, P.V. Best Practices in Perovskite Solar Cell Efficiency Measurements. Avoiding the Error of Making Bad Cells Look Good. *J. Phys. Chem. Lett.* **2015**, *6*, 852–857. [CrossRef] [PubMed]

© 2019 by the authors. Licensee MDPI, Basel, Switzerland. This article is an open access article distributed under the terms and conditions of the Creative Commons Attribution (CC BY) license (http://creativecommons.org/licenses/by/4.0/).

Article

Suitability of Copper Nitride as a Wiring Ink Sintered by Low-Energy Intense Pulsed Light Irradiation

Takashi Nakamura [1,*], Hea Jeong Cheong [2,†], Masahiko Takamura [3], Manabu Yoshida [2] and Sei Uemura [2]

1. Research Institute for Chemical Process Technology, National Institute of Advanced Industrial Science and Technology (AIST), 4-2-1 Nigatake Miyagino-ku, Sendai, Miyagi 983-8551, Japan
2. Flexible Electronics Research Center, National Institute of Advanced Industrial Science and Technology (AIST), 1-1-1 Higashi, Tsukuba, Ibaraki 305-8565, Japan; cheong.heajeong@imass.nagoya-u.ac.jp (H.J.C.); yoshida-manabu@aist.go.jp (M.Y.); sei-uemura@aist.go.jp (S.U.)
3. Nicca Chemical Co. Ltd., 23-1, 4-chome, Bunkyo, Fukui-city, Fukui 910-8670, Japan; m-takamura@niccachemical.com
* Correspondence: nakamura-mw@aist.go.jp; Tel.: +81-29-861-2272
† Present affiliation: Institute of Materials and Systems for Sustainability (IMaSS), Nagoya University, C3-1 Furo-cho Chikusa-ku, Nagoya 464-8603, Japan.

Received: 24 July 2018; Accepted: 13 August 2018; Published: 14 August 2018

Abstract: Copper nitride particles have a low decomposition temperature, they absorb light, and are oxidation-resistant, making them potentially useful for the development of novel wiring inks for printing circuit boards by means of intense pulsed light (IPL) sintering at low-energy. Here, we compared the thermal decomposition and light absorption of copper materials, including copper nitride (Cu_3N), copper(I) oxide (Cu_2O), or copper(II) oxide (CuO). Among the copper compounds examined, copper nitride had the second highest light absorbency and lowest decomposition temperature; therefore, we concluded that copper nitride was the most suitable material for producing a wiring ink that is sintered by means of IPL irradiation. Wiring inks containing copper nitride were compared with those of wiring inks containing copper nitride, copper(I) oxide, or copper(II) oxide, and copper conversion rate and sheet resistance were also determined. Under low-energy irradiation (8.3 J cm^{-2}), copper nitride was converted to copper at the highest rate among the copper materials, and provided a sheet resistance of 0.506 Ω sq^{-1}, indicating that copper nitride is indeed a candidate material for development as a wiring ink for low-energy intense pulsed light sintering-based printed circuit board production processes.

Keywords: copper; copper nitride; photo sintering; ink; paste; printed electronics; post-processing

1. Introduction

Novel methods of printing circuit boards and sensing devices are currently being developed [1–4]. Compared with traditional lithographic approaches, printing circuit boards is cost-effective and allows the production of large quantities of circuit boards with less waste. In the printing of electronic circuit boards, a conductive or semiconductive wiring ink is printed onto an insulating substrate, and the ink is then subjected to heat treatment for sintering. However, thermal treatment is problematic in that it heats not only the ink but also the substrate, which can damage the final circuit board. Therefore, alternative postprocessing methods are needed.

As alternatives to heat sintering, the application of microwaves [5], infrared light [6], and laser light [7,8] have been examined. In particular, the use of intense pulsed light (IPL) irradiation has been intensely researched [9–15]. In IPL sintering, a short pulse of light generated by a high-power xenon flash is applied to the ink and substrate; the light generated by the xenon lamp is absorbed by the ink and converted to thermal energy, which spreads throughout the circuit by thermal diffusion and causes the ink to undergo a chemical reaction. The total time required for IPL sintering is less than 1 min; therefore, IPL sintering allows printed circuit boards to be processed at high speed.

Kim's group is actively researching the use of IPL irradiation for the sintering of copper nanoparticles [9], silver nanoparticles [16], mixed silver and copper particles [17], and mixed copper particles of different sizes [18]. In their work using silver nanoparticles, they examined the temperature profile of a printed circuit pattern subjected to IPL irradiation and reported that the temperature of the ink was briefly increased to over 120 °C (the temperature needed for sintering of silver nanoparticles) without damaging the polyethylene terephthalate substrate (glass transition temperature = 62 °C). One reason why the substrate was protected from thermal damage during IPL sintering was that the ink was selectively heated because the ink and the substrate were colored and transparent, respectively. The temperature of the ink only was selectively increased for only a short time. Thus, in this respect, a suitable ink for IPL sintering is one that readily absorbs visible light.

Because of their cost-effectiveness and anti-ion-migration properties compared to silver, copper and copper-containing compounds such as copper oxide, copper salts, and organic copper complexes are useful materials for producing conductive patterns [19–22]. Although copper nanoparticles are potentially useful, they easily oxidize, and harsh thermal or reduction treatments are needed to remove the oxide layer because copper(II) oxide is chemically stable. As a novel material with which to produce a wiring ink, copper nitride (Cu_3N) may be useful because of its oxidation resistance [23] and low decomposition temperature [24]. Copper nitride has an anti-rhenium oxide structure with lattice parameters of $a = b = c = 0.3807$ nm and $\alpha = \beta = \gamma = 90°$ [25], and it decomposes to copper and nitrogen at around 400 °C [26]. Importantly, copper nitride is red-purple, which means that it absorbs visible light [27].

Here, we examined the suitability of using copper nitride for producing wiring inks sintered by using IPL irradiation. Dispersions containing copper nitride were prepared, and their light absorption and thermal decomposition properties were compared with those of inks containing copper nitride (Cu_3N), copper(I) oxide (Cu_2O), copper(II) oxide (CuO), and copper (Cu). Furthermore, two types of ink were examined: a low-viscosity liquid ink to examine the properties conferred to the ink by the copper compound, and a high-viscosity paste ink to examine a more real-world application of our ink formulations. The copper conversion ratio and sheet resistance of the inks were also determined.

2. Materials and Methods

2.1. Materials

2.1.1. Materials for Comparison with Copper Nitride

Copper(I) oxide (powder; particle size, ≤7 µm) and copper(II) oxide (nanopowder; particle size, <50 nm, as assessed by transmission electron microscopy) were purchased from Sigma-Aldrich (Tokyo, Japan). Copper nanoparticles (particle size, <100 nm) were obtained from a copper nanoparticle paste (Daiken Chemical, Osaka, Japan) by washing with acetone and then centrifuging three times.

2.1.2. Materials for Preparation of Copper Nitride Nanoparticles

Copper acetate monohydrate (special grade) and 1-nonanol (special grade) were purchased from Kanto Chemical (Tokyo, Japan). Urea (special grade) and hexane (special grade) were purchased from Wako Pure Chemical Industries (Osaka, Japan). These reagents were used without further purification.

2.1.3. Materials for Preparation of Vehicles for Liquid Inks

Ethanol (special grade) and ethylene glycol (special grade) were purchase from Wako Pure Chemical Industries (Osaka, Japan) and used without further purification.

2.1.4. Materials for Preparation of Vehicles for Paste Inks

2-(2-Butoxyethoxy)ethyl acetate (special grade), tetraethyleneglycol monomethyl ether (95%), ethyl cellulose 45 (approximately 49% ethoxy), tetrahydrofuran (special grade), hydrochloric acid (special grade), ethyl acetate (special grade), and magnesium sulfate anhydrous (special grade) were purchased from Wako Pure Chemical Industries (Osaka, Japan). (3-Glycidoxypropyl)trimethoxysilane (GPTMS, 99.2%) was purchased from Shin-Etsu Chemical (Tokyo, Japan). Tetrahydrofuran was used after distillation to remove the stabilizer that was present in the purchase solution; the other reagents were used without further purification.

2.2. Preparation of Copper Nitride Nanoparticles

Copper nitride nanoparticles were prepared in accordance with previous methods [27]. Copper acetate monohydrate (20 mmol) and urea (100 mmol) were mixed with 1-nonanol (400 mL) in a three-neck flask. The atmosphere in the flask was replaced with nitrogen. The solution was heated by microwave irradiation at 190 °C for 60 min (ramp, 30 min; hold, 30 min) by using a microwave oven (μReactorEx, Shikoku Keisoku, Kagawa, Japan) equipped with a fiber optic thermometer (Anritsu Keiki, Tokyo, Japan) to obtain a liquid suspension. After the suspension was centrifuged (30,000× g for 60 min) and washed with hexane three times, a red-purple powder was obtained.

2.3. Preparation of Liquid Inks

A sample of one of the test compounds (0.15 mg) was mixed with ethanol (1 mL) in a microcentrifuge tube, and the mixture was milled by using a homogenization pestle for 5 min. Inks containing a reductant were prepared in the same way but with the addition of ethylene glycol (0.2 mL). Aliquots (0.2 mL) of the liquid inks were dropped onto a cover glass by using a micropipette and then dried in an oven at 60 °C for 2 h (Figure S1a).

2.4. Synthesis of Poly(3-glycidoxypropyl)trimethoxysilane as an Adhesion Compound

Poly(3-glycidoxypropyl)trimethoxysilane (PGPTMS) was synthesized in accordance with previous methods. [28,29] Briefly, GPTMS (24.8 g), distilled water (5.4 g), and aqueous hydrochloric acid (0.1 N, 5 mL) were mixed with tetrahydrofuran (200 mL) in a flask attached to a condenser. The solution was heated under reflux for 8 h, diluted with ethyl acetate (300 mL), and then washed with distilled water three times. The organic phase was dried over anhydrous magnesium sulfate for 2 h and then filtered. A yellow viscous liquid was obtained after evaporation of the organic solvent.

2.5. Preparation of Paste Inks

The vehicle for the paste inks was typically prepared as follows: In a disposable plastic cup, 2-(2-butoxyethoxy)ethyl acetate (41.85 g), diethylene glycol monobutyl ether (4.65 g), ethyl cellulose (3.5 g), and PGPTMS (0.5 g) were combined and then transferred to a planetary centrifugal mixer (ARE-310; THINKY, Tokyo, Japan) and mixed at 2000 rpm for 10 min. To prepare the paste inks, each of the copper-containing materials was mixed with vehicle in a microcentrifuge tube at a weight ratio of 1:1, and the mixture was then milled by using a homogenization pestle for 5 min.

To prepare films of the paste inks, two strips of adhesive tape (thickness, 0.058 mm) were placed across a cover glass (18 × 18 mm) at an interval of 10 mm. An aliquot (approximately 0.1 mL) of paste ink was dropped onto the cover glass in the space between the two strips of adhesive tape and spread by using a glass rod (Figure S1b). The obtained ink film was dried in an oven at 60 °C for 2 h. The thickness of the prepared ink film was the thickness of the tape.

2.6. Intense Pulsed Light Sintering

Induced pulsed light sintering was performed by using a high-power xenon flash lamp system (SINTERON 2010-L; Xenon, MA, USA) equipped with a type B lamp. The wavelength range of the emitted light was 240–1000 nm. The experimental conditions used for the IPL sintering are indicated where necessary in the text.

2.7. Characterization of Copper Compounds and Inks

Diffuse reflection spectra were obtained by using a UV-3150 spectrometer (Shimadzu, Kyoto, Japan) attached to an integrating sphere (ISR-260; Shimadzu). The Kubelka–Munk equation was used to estimate the absorption ratio of the samples from the obtained reflectance. Thermogravimetric-differential thermal analysis was carried out by using a Thermo plus Evo2 instrument (Rigaku, Tokyo, Japan); samples were heated from 25 to 500 °C at 10 °C/min under air or an inert atmosphere of nitrogen, and alumina powder was used as the reference material. The crystal phase of samples was determined by means of powder X-ray diffraction analysis by using a SmartLab instrument (Rigaku) with experimental conditions of 40 kV and 30 mA from $2\theta = 10°$ to $90°$. To measure the X-ray diffraction of films prepared from the inks, the films attached with the substrate were directly placed on a zero-background holder. The morphology of the films was examined by means of scanning electron microscopy by using a TM-1000 (Hitachi, Tokyo, Japan) or S-4800 instrument (Hitachi, Tokyo, Japan) at an electron acceleration voltage of 15 kV (TM-1000) or 3 kV (S-4800), respectively. Samples were observed without a conductive coating. The sheet resistance of the films was measured by using a Loresta-GP resistivity meter (MCP-T610; Mitsubishi Chemical Analytech, Kanagawa, Japan) equipped with a four-point probe (PSP; Mitsubishi Chemical Analytech).

2.8. Calculation of Copper Conversion Ratio

Copper conversion ratios were estimated as follows: First, diffraction patterns obtained by means of X-ray diffraction analysis for the sintered films were fitted by using a pseudo Voigt function for peak separation. The peak separation data were then compared with reported patterns for Cu_3N (PDF2 no. 1-73-6209), Cu (PDF2 no. 3-65-9026), CuO (PDF2 no. 1-89-5896), and Cu_2O (PDF2 no. 1-78-2076) contained in the International Centre for Diffraction database. Next, the integral strength of the crystal planes for each crystal phase was estimated. The series of operations described above was performed by using the PDXL integrated X-ray powder diffraction software (Rigaku). Finally, the integral strengths and the following equation were used to calculate the copper conversion ratio:

$$R_{Cu\ conver.} = \frac{I_{Cu}}{I_{Cu3N} + I_{Cu} + I_{CuO} + I_{Cu2O}}$$

where I_{Cu3N}, I_{Cu}, I_{CuO}, and I_{Cu2O} are the integral strengths for (111) of copper nitride ($2\theta = 40.9°$), copper ($2\theta = 43.3°$), copper(II) oxide ($2\theta = 38.7°$), and copper(I) oxide ($2\theta = 36.4°$), respectively.

3. Results and Discussion

3.1. Comparison of the Light Absorption and Thermal Decomposition Properties of the Copper Compounds

We first obtained diffuse reflectance spectra of the copper compounds and found that diffuse reflectance increased in the order $CuO > Cu_3N > Cu_2O > Cu$ (Figure 1). We then examined the thermal decomposition of Cu_3N, CuO, and Cu_2O under an air or N_2 atmosphere (Figure 2). Under the N_2 atmosphere, the weight of Cu_3N was decreased at 100 °C, and at 250 and 500 °C the weight losses were −5.4% and −5.1%, respectively. In contrast, the weights of CuO and Cu_2O did not change until 500 °C, at which point the weights had increased by 0.1 and 1.1%, respectively. Under the air atmosphere, the weight of copper nitride decreased until 188 °C, after which, it increased via oxidation; at 500 °C, the weight change of copper nitride was −0.7%. The weight of Cu_2O remained constant until 333 °C at which point it began to increase as a result of oxidation; the weight was 7.2% at 500 °C. The weight of CuO remained constant, and the change was only 0.3% at 500 °C. Thus, Cu_3N decomposed at a lower temperature than CuO and Cu_2O, which is consistent with previous reports that CuO and Cu_2O melt at 1193 and 1229 °C [30]. The light absorption and thermal decomposition properties of the materials are summarized in Table 1. Among the copper compounds examined, copper nitride had the second highest light absorbency and lowest decomposition temperature; therefore, we concluded that copper nitride was the most suitable material for producing a wiring ink that is sintered by means of IPL irradiation.

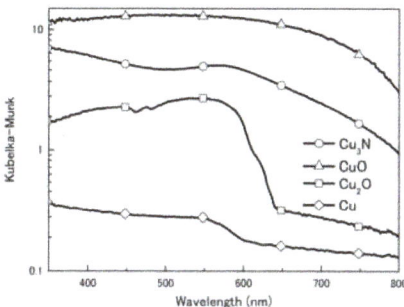

Figure 1. Ultraviolet–visible diffuse reflection spectra of copper nitride (Cu_3N), copper(I) oxide (Cu_2O), copper(II) oxide (CuO), and copper (Cu) particles.

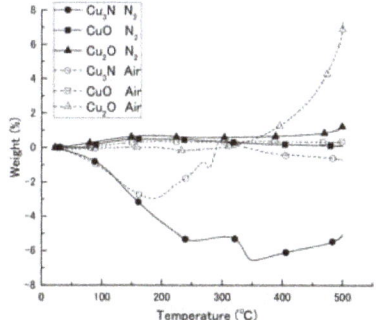

Figure 2. Thermogravimetric-differential thermal analysis curves of copper nitride (Cu_3N), copper(I) oxide (Cu_2O), and copper(II) oxide (CuO) in an air or N_2 atmosphere.

Table 1. Chemical properties of the copper-containing materials.

Copper-Containing Material	Light Absorbency		Decomposition Temperature or Melting Point (°C)	Copper Content Ratio
	Absorption Range (nm)	Strength		
Cu_3N	350–800	Medium	250	93
CuO	350–800	Strong	1193	87
Cu_2O	350–650	Medium	1229	89
Cu	350–800	Weak	1085	100

3.2. Evaluation of Liquid Inks

Simple liquid dispersions containing the copper compounds were prepared. To increase the reduction property of the copper compounds, liquid dispersions containing ethylene glycol as a reduction compound were also prepared. Films of the inks were prepared by dropping the ink onto a cover glass, and IPL sintering was performed at 12.45 and 16.60 J cm^{-2} with pulse widths of 1500 and 2000 µs, respectively; for both irradiation conditions, the applied voltage, irradiation period, and number of pulses were 2.3 kV, 1000 ms, and 1, respectively. Table 2 shows the copper conversion ratios, as estimated from the integral strength obtained by means of X-ray diffraction (Figure S2), and sheet resistances of the films made from the liquid dispersions.

In the absence of ethylene glycol, copper nitride film irradiated at 12.45 and 16.60 J cm^{-2} had copper conversion ratios of 0.91 and 0.88, respectively. However, in the presence of ethylene glycol, copper nitride film irradiated at 12.45 and 16.60 J cm^{-2} had copper conversion ratios of 0.99 and 0.96, respectively. No conversion to copper was observed for the copper(I) oxide and copper(II) oxide films, irrespective of the absence or presence of ethylene glycol.

Sheet resistance after IPL sintering was measured by means of a four-point probe method in the presence or absence of ethylene glycol (Table 2 and Figure 3). The sheet resistances for the copper(I) oxide films were beyond the maximum limit of quantitation of the instrument (>9.999 × 10^7 Ω). Similarly, sheet resistances for the copper(II) oxide films could not be obtained because the films broke apart after sintering (Figure S3).

Table 2. Conversion ratio and sheet resistance obtained with different irradiation energy densities and vehicles.

Energy Density (J cm^{-2})	Vehicle	Conversion Ratio			Sheet Resistance (Ω sq^{-1})			
		Cu$_3$N	CuO	Cu$_2$O	Cu$_3$N	CuO	Cu$_2$O	Cu
12.45	EtOH	0.91	0	0	4.52 × 10^6	N.D.	O.L.	2.70 × 10^{-1}
16.60	EtOH	0.88	0	0	2.37 × 10^0	N.D.	O.L.	1.11 × 10^{-1}
12.45	EtOH + EG	0.99	0	0	1.34 × 10^0	N.D.	O.L.	3.31 × 10^{-1}
16.60	EtOH + EG	0.96	0	0	6.95 × 10^{-1}	N.D.	O.L.	4.72 × 10^{-1}

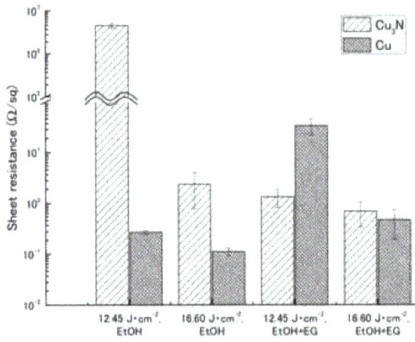

Figure 3. Sheet resistance of films containing copper nitride (Cu$_3$N) and copper (Cu) after intense pulsed light sintering. Vehicle, ethanol (EtOH) with or without ethylene glycol (EG).

For the copper nitride films, higher sheet resistances were obtained in the absence of ethylene glycol (4.52 × 10^6 and 2.37 × 10^0 Ω sq^{-1} at 12.45 and 16.60 J cm^{-2}, respectively) than in the presence of ethylene glycol (1.34 × 10^0 and 6.95 × 10^{-1} Ω sq^{-1} at 12.45 and 16.60 J cm^{-2}, respectively). However, the reason for the particularly high sheet resistance obtained for the film not containing ethylene glycol and irradiated at 12.45 J cm^{-2} was that the conductive path in the film was broken by the probes during measurement, because the mechanical strength of the film was weak; visual inspection after measurement of the sheet resistance revealed that the film was broken. Compared with the sheet resistances of the films containing the copper nanoparticles, the sheet resistances of the copper nitride films were comparable or one order of magnitude higher.

Visual and scanning electron microscopy (Figure 4) inspection revealed that the surface of the film containing ethylene glycol became coarse after IPL irradiation at 16.60 J cm^{-2}. In addition, the surface was found to contain hollow particles of copper with a particle size of 5 μm or more. The coarse particle size and failure of necking to form between particles are likely the reasons for the weakness of the film. We propose the following process for the formation of the hollow copper particles (Figure S4): Copper nitride particles are chemically changed to copper particles by the thermal energy produced by IPL irradiation; these particles rapidly fuse with nearby particles because of the decreased surface energy of the particles, and copper plate-like films form and then become round owing to surface tension. This would suggest that it is necessary to avoid rapid heating by IPL irradiation [16] and rapid cooling of the particles to prevent the generation of hollow particles. Specifically, it relaxes the heating rate to transfer the thermal energy, which is rapidly generated by IPL irradiation, to organic compound. As a result, we conclude that the condition of the film could be improved because the intense reaction is inhibited.

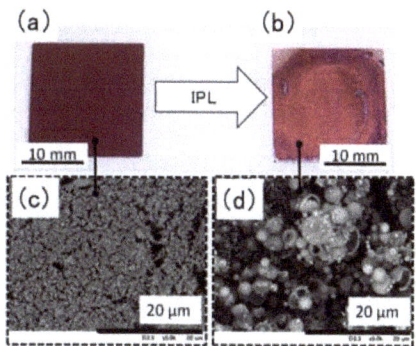

Figure 4. Appearance and scanning election microscopy images of a film containing ethylene glycol (a,c) before and (b,d) after intense pulsed light irradiation at 16.60 J cm^{-2}.

3.3. Evaluation of Paste Inks

3.3.1. Use of Ethyl Cellulose and PGPTMS as a Binder and an Adhesion Reagent, respectively, to Produce a Copper Nitride Paste Ink

Although copper nitride was found to have a high copper conversion ratio, some of the films produced had a high sheet resistance because of weak mechanical strength. To address this issue, and to examine a more practical formulation for copper nitride ink, we investigated the use of a paste ink in which ethyl cellulose and poly(3-glycidoxypropyl)trimethoxysilane (PGPTMS) were used as a binder and adhesion reagent, respectively. It was previously reported that PGPTMS improves bonding strength between the particles in ink films and between the substrate and the film [28,29].

First, we determined the most suitable concentration of PGPTMS to use in the vehicle by preparing vehicles containing different concentrations of PGPTMS and using them to make films containing copper nitride at a weight ratio of 1:1 (Table 3). The films were then exposed to IPL irradiation under different conditions (Table 4), and copper conversion ratio and sheet resistance were measured (Table 5, Figure 5, and Figure S5).

Table 3. Vehicle compositions.

Type of Vehicle	PGPTMS (wt %)	2-(2-Butoxyethoxy)ethyl acetate (wt %)	Diethylene Glycol Monobutyl Ether (wt %)	Ethyl Cellulose (wt %)
Vehicle 1	0	83.7	9.3	7
Vehicle 2	1	82.9	9.2	6.9
Vehicle 3	7	77.9	8.7	6.5

Table 4. Intense pulsed light sintering conditions.

Photo Sintering Condition	Applied Voltage (kV)	Pulse Width (µs)	Period (ms)	Number of Pulses	Total Electrical Energy (J)	Energy Density (J cm^{-2})	Distance (mm)
P.S.1	2.0	1000	1000	1	344.0	5.94	25
P.S.2	2.3	1000	1000	1	481.0	8.30	25
P.S.3	2.3	1500	1000	1	721.6	12.45	25
P.S.4	2.3	2000	1000	1	962.1	16.60	25
P.S.5	2.3	1000	1000	4	481.0	8.30	25

Table 5. Copper conversion ratio and sheet resistance obtained by using the intense pulsed light sintering conditions shown in Table 4.

IPL Sintering Condition	Conversion Ratio					Sheet Resistance (Ω sq^{-1})				
	Cu$_3$N			CuO	Cu$_2$O	Cu$_3$N			CuO	Cu$_2$O
Vehicle	1	2	3	2	2	1	2	3	2	2
P.S.1	0.10	0.43	0.35	0.00	0.00	O.L.	2.74 × 10^0	O.L.	O.L.	O.L.
P.S.2	0.41	0.63	0.68	0.00	0.00	O.L.	5.06 × 10^{-1}	O.L.	O.L.	O.L.
P.S.3	0.62	0.84	0.67	0.03	0.00	9.65 × 10^0	1.19 × 10^0	9.65 × 10^0	O.L.	O.L.
P.S.4	0.69	0.82	0.62	0.10	0.00	1.41 × 10^0	7.90 × 10^0	2.82 × 10^0	O.L.	O.L.
P.S.5	0.46	0.75	0.56	0.00	0.00	O.L.	4.97 × 10^{-1}	O.L.	O.L.	O.L.

O.L. = overload.

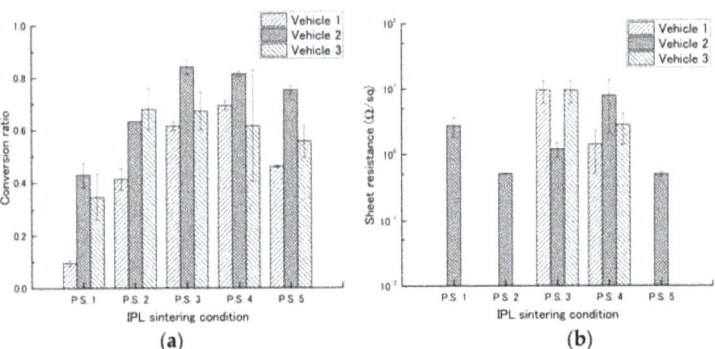

Figure 5. Conversion ratio (a) and sheet resistance (b) of films made with various vehicles (Table 3) and then exposed to intense pulsed light irradiation using the conditions shown in Table 4.

Among the inks containing vehicles 1, 2, and 3 (PGPTMS concentrations, 0, 1, and 7 wt %, respectively), the ink containing vehicle 2 had a high copper conversion rate under most conditions. We therefore examined the effect of different IPL irradiation conditions on this ink. In a comparison of different applied voltages (2.0 [P.S.1] or 2.3 kV [P.S.2]), the copper conversion rate was increased from 0.43 to 0.63 by increasing the applied voltage. In a comparison of different pulse widths (1000 [P.S.2], 1500 [P.S.3], or 2000 µs [P.S.4]), the copper conversion rate increased with increasing pulse width, although the copper conversion rate was comparable between pulse widths of 1500 and 2000 µs. In a comparison of number of pulses (1 [P.S.2] or 4 pulses [P.S.5]), the copper conversion rate was increased from 0.63 to 0.75 by increasing the number of pulses. In addition, low sheet resistances, 5.06 × 10^{-1} and 4.97 × 10^{-1} Ω sq^{-1}, were obtained for conditions P.S.2 and P.S.5, respectively. No relationship was observed between copper conversion rate and sheet resistance.

To examine the relationship between sheet resistance and bonding between particles, the surface condition of the sintered samples prepared by using vehicle 2 was observed by means of scanning electron microscopy (Figure 6). Improved necking and no increase in particle size above micrometer size were observed in the film irradiated using the P.S.2 conditions (8.30 J cm^{-2}). In addition, the influence of the additives (i.e., ethyl cellulose and PGPTMS) could be clearly observed between the paste inks (Figure 6) and the liquid inks (Figure 4). The additives inhibited the intense reaction of copper nitride by IPL irradiation (described above in the Section 3.2) because the heat energy produced by IPL irradiation spread not only to the copper nitride, but also to the additives. As a result, the formation of the particles above micrometer size with a hollow shape was inhibited. Furthermore, it appears that the necking between particles was improved by the addition of the additives because the sheet resistance was decreased and mechanical strength was increased in films containing the additives compared with those without.

In this series of experiments, the copper conversion ratio was limited to approximately 0.8 for the paste ink, owing to the penetration depth of the IPL irradiation. For all of the examined films, there was a difference in color after IPL irradiation when both sides of the film were compared (Figure S6); the back of the film was darker than the front, indicating that the copper nitride was not converted to copper at the back of the film.

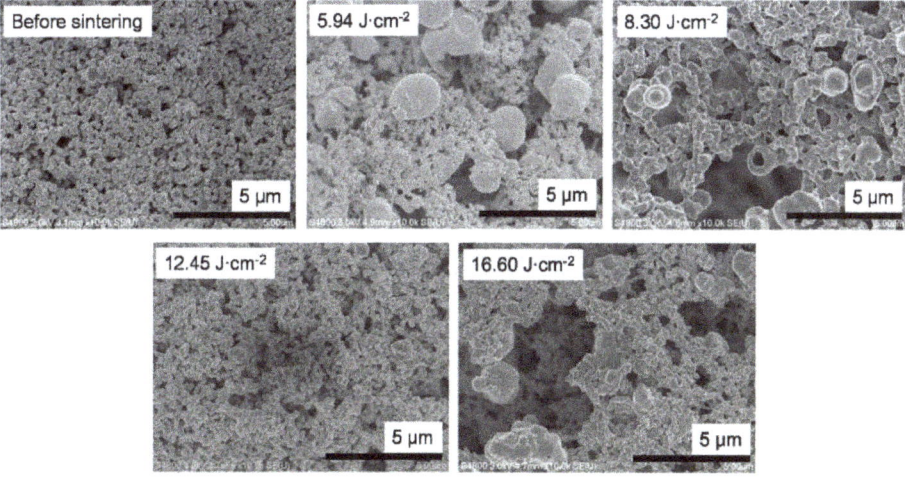

Figure 6. Scanning electron microscopy images of films prepared by using vehicle 2 before and after intense pulsed light sintering with different irradiation conditions.

3.3.2. Comparison of Paste Inks Prepared with Vehicle 2

Paste inks containing copper nitride, copper(II) oxide, or copper(I) oxide were prepared with vehicle 2 and subjected to IPL sintering (sintering conditions, Table 4; copper conversion ratio and sheet resistance, Table 5; X-ray diffraction patterns, Figure S7). The appearances of the samples before and after IPL irradiation are shown in Figure 7. For the copper nitride films, the surface color was clearly changed from dark to light brown. In contrast, the color of the copper(II) oxide and copper(I) oxide films remained largely unchanged (minor changes were observed for copper(II) oxide under conditions P.S.3 and P.S.4).

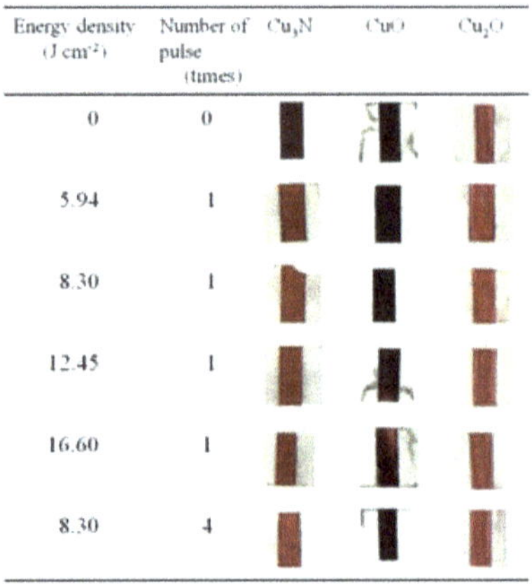

Energy density (J cm^{-2})	Number of pulse (times)	Cu$_3$N	CuO	Cu$_2$O
0	0			
5.94	1			
8.30	1			
12.45	1			
16.60	1			
8.30	4			

Figure 7. Appearance of films prepared with vehicle 2 and containing (Cu$_3$N), copper(I) oxide (Cu$_2$O), or copper(II) oxide (CuO) before and after intense pulsed light irradiation under the conditions shown in Table 4.

The sheet resistances of the copper(II) oxide and copper(I) oxide films could not be determined because they were beyond the maximum limit of quantitation of the instrument (>9.999 × 10^7 Ω). Thus, in the case of copper(I) oxide film, sheet resistance was not improved despite the film having improved mechanical strength compared to the film prepared by liquid ink.

3.4. Volume Resistivity and Comparison with that of Previous Studies

The volume resistivity of the obtained conductive film prepared by using the vehicle 2 after IPL irradiation at 8.30 J cm^{-2} was estimated. The thickness of the film after IPL irradiation was 20 µm, as determined from a cross-sectional scanning electron microscopy image (Figure S8). Assuming that 60% (Table 5) of the film from the sample surface was converted from copper nitride to copper, it is estimated that a conductive layer with a thickness of 12 µm was produced.

The results of previously published studies are shown in Table 6 together with the results of the present study. In these previously published studies, various materials (i.e., copper, copper(II) oxide, and copper organic complex) and particle shapes (i.e., nanoparticles, microparticles, and nanowire) were successfully used to produce conductive films via IPL irradiation. Although direct comparison of the results of these studies is difficult owing to differences in the spectra of the light sources used, the film in the present study had lower resistivity obtained by IPL irradiation with lower energy density than those previously reported. In the future, we intend to find the irradiation conditions and additives to obtain a conductive film with better characteristics.

Table 6. Comparison of compound information and obtained values of resistance described in the present study with previously reported values.

Entry	Compound	Form	Lower IPL Irradiation Energy			Higher IPL Irradiation Energy			References
			Energy Density (J cm^{-2})	Volume Resistivity ($\mu\Omega$ cm)	Sheet Resistance (mΩ sq^{-1})	Energy Density (J cm^{-2})	Volume Resistivity ($\mu\Omega$ cm)	Sheet Resistance (mΩ sq^{-1})	
1	Copper nitride	Nanoparticle	8.3	607	506	–	–	–	This work
2	Copper	Nanoparticle	20	Infi.	Infi.	45	5	–	[9]
3	Copper	Nanoparticle	12	Infi.	Infi.	32	173	–	[12]
4	Copper	Nanoparticle	10	–	9860	12.5	–	72	[31]
5	Copper	Nanoparticle	8.0	450	–	12.5	54	–	[32]
6	Copper	Nanoparticle + microparticle	10.0	438	–	12.5	80	–	[18]
7	Copper	Nanoparticle + nanowire	7.5	420–520	–	12.5	22.7	–	[33]
8	Copper(II) oxide	–	10.9	–	200	11	–	170	[34]
9	Copper organic complex	–	16.5	750	–	40.6	4.6	–	[11]

Infi. = infinite.

4. Conclusions

Here, we examined the properties of films containing copper nitride, copper(II) oxide, and copper(I) oxide for their suitability as wiring inks for IPL sintering-based printed circuit board production. The following was clarified: Among the copper compounds examined, copper nitride had the second highest light absorption and lowest decomposition temperature, suggesting that it is suitable for sintering by IPL irradiation. Liquid ink containing copper nitride had a copper conversion ratio of 0.96 at an irradiation energy of 16.6 J cm^{-2}. The sheet resistance of the film was on the order of 10^{-1} Ω sq^{-1}, which was comparable to that of a film made from liquid ink containing copper nanoparticles. To improve the mechanical strength of the film, paste inks containing ethyl cellulose and PGPTMS as a binder and adhesion compound, respectively, were prepared. The optimum amount of PGPTMS was 1 wt %, and a film made from a paste ink containing the optimized vehicle and copper nitride and irradiated at 8.30 J cm^{-2} had a sheet resistance of 5.06×10^{-1} Ω sq^{-1}. Together, the present results indicate that copper nitride is a suitable material for the development of wiring inks sintered by means of IPL irradiation.

Supplementary Materials: The following are available online at http://www.mdpi.com/2079-4991/8/8/617/s1, Figure S1: schematic images for preparing test film of (a) liquid ink and (b) paste ink, Figure S2: XRD patterns of samples for copper nitride, copper(II) oxide, and copper(I) oxide after IPL sintering, Figure S3: appearance of samples prepared from liquid ink including CuO before and after IPL sintering, Figure S4: schematic image for forming hollow particles, Figure S5: XRD patterns of different vehicles including Cu$_3$N films after IPL sintering, Figure S6: appearance of a sample film after IPL sintering front side and back side, Figure S7: XRD patterns of sample films including CuO and Cu$_2$O after IPL sintering, Figure S8: cross-section SEM image of the sample sintered by IPL at 8.30 J cm^{-2} of irradiation energy.

Author Contributions: Conceptualization, T.N.; Methodology, T.N., S.U. and M.Y.; Validation, T.N. and H.J.C.; Formal Analysis, T.N. and H.J.C.; Investigation, T.N., H.J.C. and S.U.; Resources, T.N., H.J.C., M.T., S.U. and M.Y.; Data Curation, T.N. and H.J.C.; Writing—Original Draft Preparation, T.N.; Writing—Review & Editing, T.N.; Visualization, T.N.; Supervision, T.N.; Project Administration, T.N. and M.T.; Funding Acquisition, T.N. and M.T.

Funding: This work was financially supported by a grant from the Adaptable and Seamless Technology Transfer Program (A–STEP) through the Target-Driven R&D Program of the Japan Science and Technology Agency, JST (no. AS2621364M).

Acknowledgments: This research is supported by A–STEP of JST.

Conflicts of Interest: The authors have no conflicts of interest directly relevant to the content of this article.

References

1. Wünscher, S.; Abbel, R.; Perelaer, J.; Schubert, U.S. Progress of alternative sintering approaches of inkjet-printed metal inks and their application for manufacturing of flexible electronic devices. *J. Mater. Chem. C* **2014**, *2*, 10232–10261. [CrossRef]
2. Perelaer, J.; Smith, P.J.; Mager, D.; Soltman, D.; Volkman, S.K.; Subramanian, V.; Korvink, J.G.; Schubert, U.S. Printed electronics: The challenges involved in printing devices, interconnects, and contacts based on inorganic materials. *J. Mater. Chem.* **2010**, *20*, 8446–8453. [CrossRef]
3. Perelaer, J.; Schubert, U.S. Novel approaches for low temperature sintering of inkjet-printed inorganic nanoparticles for roll-to-roll (R2R) applications. *J. Mater. Res.* **2013**, *28*, 564–573. [CrossRef]
4. Singh, R.; Singh, E.; Nalwa, H.S. Inkjet printed nanomaterial based flexible radio frequency identification (RFID) tag sensors for the internet of nano things. *RSC Adv.* **2017**, *7*, 48597–48630. [CrossRef]
5. Perelaer, J.; Jani, R.; Grouchko, M.; Kamyshny, A.; Magdassi, S.; Schubert, U.S. Plasma and microwave flash sintering of a tailored silver nanoparticle ink, yielding 60% bulk conductivity on cost- effective polymer foils. *Adv. Mater.* **2012**, *24*, 3993–3998. [CrossRef] [PubMed]
6. Hwang, H.J.; Oh, K.H.; Kim, H.S. All-photonic drying and sintering process via flash white light combined with deep-UV and near-infrared irradiation for highly conductive copper nano-ink. *Sci. Rep.* **2016**, *6*, 19696. [CrossRef] [PubMed]
7. Zhao, W.; Rovere, T.; Weerawarne, D.; Osterhoudt, G.; Kang, N.; Joseph, P.; Luo, J.; Shim, B.; Poliks, M.; Zhong, C.-J. Nanoalloy printed and pulse-laser sintered flexible sensor devices with enhanced stability and materials compatibility. *ACS Nano* **2015**, *9*, 6168–6177. [CrossRef] [PubMed]

8. Zenou, M.; Ermak, O.; Saar, A.; Kotler, Z. Laser sintering of copper nanoparticles. *J. Phys. D* **2014**, *47*, 25501. [CrossRef]
9. Kim, H.S.; Dhage, S.R.; Shim, D.E.; Hahn, H.T. Intense pulsed light sintering of copper nanoink for printed electronics. *Appl. Phys. A* **2009**, *97*, 791–798. [CrossRef]
10. Dharmadasa, R.; Jha, M.; Amos, D.A.; Druffel, T. Room temperature synthesis of a copper ink for the intense pulsed light sintering of conductive copper films. *ACS Appl. Mater. Interfaces* **2013**, *5*, 13227–13234. [CrossRef] [PubMed]
11. Wang, B.Y.; Yoo, T.H.; Song, Y.W.; Lim, D.S.; Oh, Y.J. Cu ion ink for a flexible substrate and highly conductive patterning by intensive pulsed light sintering. *ACS Appl. Mater. Interfaces* **2013**, *5*, 4113–4119. [CrossRef] [PubMed]
12. Han, W.S.; Hong, J.M.; Kim, H.S.; Song, Y.W. Multi-pulsed white light sintering of printed Cu nanoinks. *Nanotechnology* **2011**, *22*, 395705. [CrossRef] [PubMed]
13. Jha, M.; Dharmadasa, R.; Draper, G.L.; Sherehiy, A.; Sumanasekera, G.; Amos, D.; Druffel, T. Solution phase synthesis and intense pulsed light sintering and reduction of a copper oxide ink with an encapsulating nickel oxide barrier. *Nanotechnology* **2015**, *26*, 175601. [CrossRef] [PubMed]
14. Albrecht, A.; Rivadeneyra, A.; Abdellah, A.; Lugli, P.; Salmerón, J.F. Inkjet printing and photonic sintering of silver and copper oxide nanoparticles for ultra-low-cost conductive patterns. *J. Mater. Chem. C* **2016**, *4*, 3546–3554. [CrossRef]
15. Kang, H.; Sowade, E.; Baumann, R.R. Direct intense pulsed light sintering of inkjet-printed copper oxide layers within six milliseconds. *ACS Appl. Mater. Interfaces* **2014**, *6*, 1682–1687. [CrossRef] [PubMed]
16. Chung, W.H.; Hwang, H.J.; Lee, S.H.; Kim, H.S. In situ monitoring of a flash light sintering process using silver nano-ink for producing flexible electronics. *Nanotechnology* **2013**, *24*, 35202. [CrossRef] [PubMed]
17. Chung, W.H.; Hwang, Y.T.; Lee, S.H.; Kim, H.S. Electrical wire explosion process of copper/silver hybrid nano-particle ink and its sintering via flash white light to achieve high electrical conductivity. *Nanotechnology* **2016**, *27*, 205704. [CrossRef] [PubMed]
18. Joo, S.J.; Hwang, H.J.; Kim, H.S. Highly conductive copper nano/microparticles ink via flash light sintering for printed electronics. *Nanotechnology* **2014**, *25*, 265601. [CrossRef] [PubMed]
19. Li, Y.; Qi, T.; Chen, M.; Xiao, F. Mixed ink of copper nanoparticles and copper formate complex with low sintering temperatures. *J. Mater. Sci.* **2016**, *27*, 11432–11438. [CrossRef]
20. Yang, W.D.; Liu, C.Y.; Zhang, Z.Y.; Liu, Y.; Nie, S.D. Copper inks formed using short carbon chain organic Cu-precursors. *RSC Adv.* **2014**, *4*, 60144–60147. [CrossRef]
21. Yang, W.D.; Wang, C.H.; Arrighi, V.; Liu, C.Y.; Watson, D. Microstructure and electrical property of copper films on a flexible substrate formed by an organic ink with 9.6% of Cu content. *J. Mater. Sci.* **2015**, *26*, 8973–8982. [CrossRef]
22. Santos, G.M.C. Cu_2O Nanoparticles for Application in Printed and Flexible Electronics. Ph.D. Thesis, Universidade Nova de Lisboa, Lisbon, Portugal, 2017.
23. Nakamura, T.; Katayama, M.; Watanabe, T.; Inada, Y.; Ebina, T.; Yamaguchi, A. Stability of copper nitride nanoparticles under high humidity and in solutions with different acidity. *Chem. Lett.* **2015**, *44*, 755–757. [CrossRef]
24. Nakamura, T.; Hayashi, H.; Hanaoka, T.; Ebina, T. Preparation of copper nitride (Cu_3N) nanoparticles in long-chain alcohols at 130–200 °C and nitridation mechanism. *Inorg. Chem.* **2014**, *53*, 710–715. [CrossRef] [PubMed]
25. Juza, R.; Hahn, H. Crystal structures of Cu_3N, GaN and InN-metallic amides and metallic nitrides. *Z. Anorg. Allg. Chem.* **1938**, *239*, 282–287. [CrossRef]
26. Fan, X.Y.; Li, Z.J.; Meng, A.; Li, C.; Wu, Z.G.; Yan, P.X. Improving the thermal stability of Cu_3N films by addition of mn. *J. Mater. Sci. Technol.* **2015**, *31*, 822–827. [CrossRef]
27. Nakamura, T.; Hayashi, H.; Ebina, T. Preparation of copper nitride nanoparticles using urea as a nitrogen source in a long-chain alcohol. *J. Nanopart. Res.* **2014**, *16*, 2699. [CrossRef]
28. Hong, J.U.; Kumar, A.; Han, H.S.; Koo, Y.H.; Kim, H.W.; Park, J.H.; Kang, H.S.; Lee, B.C.; Piao, L.; Kim, S.H. Alkoxysilane adhesion promoter for Ag nano-ink. *Bull. Korean Chem. Soc.* **2013**, *34*, 2539–2542. [CrossRef]
29. Jiang, J.; Koo, Y.H.; Kim, H.W.; Park, J.H.; Kang, H.S.; Lee, B.C.; Kim, S.H.; Song, H.E.; Piao, L. High-temperature adhesion promoter based on (3-glycidoxypropyl) trimethoxysilane for Cu paste. *Bull. Korean Chem. Soc.* **2014**, *35*, 3025–3029. [CrossRef]

30. Predel, B. Cu–O (copper-oxygen). In *Cr–Cs–Cu–Zr*; Madelung, O., Ed.; Springer: Berlin/Heidelberg, Germany, 1994; Volume 5D.
31. Hwang, H.J.; Chung, W.H.; Kim, H.S. In situ monitoring of flash-light sintering of copper nanoparticle ink for printed electronics. *Nanotechnology* **2012**, *23*, 485205. [CrossRef] [PubMed]
32. Kim, Y.J.; Ryu, C.H.; Park, S.H.; Kim, H.S. The effect of poly (N-vinylpyrrolidone) molecular weight on flash light sintering of copper nanopaste. *Thin Solid Films* **2014**, *570*, 114–122. [CrossRef]
33. Joo, S.J.; Park, S.H.; Moon, C.J.; Kim, H.S. A highly reliable copper nanowire/nanoparticle ink pattern with high conductivity on flexible substrate prepared via a flash light-sintering technique. *ACS Appl. Mater. Interfaces* **2015**, *7*, 5674–5684. [CrossRef] [PubMed]
34. Paquet, C.; James, R.; Kell, A.J.; Mozenson, O.; Ferrigno, J.; Lafreniere, S.; Malenfant, P.R.L. Photosintering and electrical performance of CuO nanoparticle inks. *Org. Electron.* **2014**, *15*, 1836–1842. [CrossRef]

© 2018 by the authors. Licensee MDPI, Basel, Switzerland. This article is an open access article distributed under the terms and conditions of the Creative Commons Attribution (CC BY) license (http://creativecommons.org/licenses/by/4.0/).

Article

Investigation on the Selenization Treatment of Kesterite $Cu_2Mg_{0.2}Zn_{0.8}Sn(S,Se)_4$ Films for Solar Cell

Dongyue Jiang [1], Yu Zhang [1], Yingrui Sui [1,*], Wenjie He [1], Zhanwu Wang [1], Lili Yang [1], Fengyou Wang [1] and Bin Yao [2]

1. Key Laboratory of Functional Materials Physics and Chemistry of the Ministry of Education, Jilin Normal University, Jilin 136000, China
2. State Key Laboratory of Superhard Materials and College of Physics, Jilin University, Jilin 130012, China
* Correspondence: syr@jlnu.edu.cn; Tel.: +86-43-4329-4566

Received: 27 May 2019; Accepted: 21 June 2019; Published: 29 June 2019

Abstract: High-selenium $Cu_2Mg_{0.2}Zn_{0.8}Sn(S,Se)_4$ (CMZTSSe) films were prepared on a soda lime glass substrate using the sol–gel spin coating method, followed by selenization treatment. In this work, we investigated the effects of selenization temperature and selenization time on the crystal quality, and electrical and optical properties of CMZTSSe films. The study on the micro-structure by XRD, Raman, X-ray photoelectron spectroscopy (XPS), and energy-dispersive X-ray spectroscopy (EDS) analysis showed that all CMZTSSe samples had kesterite crystalline structure. In addition, the crystalline quality of CMZTSSe is improved and larger Se takes the site of S in CMZTSSe with the increase of selenization temperature and selenization time. When increasing the selenization temperature from 500 to 530 °C and increasing the annealing time from 10 to 15 min, the morphological studies showed that the microstructures of the films were dense and void-free. When further increasing the temperature and time, the crystalline quality of the films began to deteriorate. In addition, the bandgaps of CMZTSSe are tuned from 1.06 to 0.93 eV through adjusting the selenization conditions. When CMZTSSe samples are annealed at 530 °C for 15 min under Se atmosphere, the crystal quality and optical–electrical characteristics of CMZTSSe will be optimal, and the grain size and carrier concentration reach maximums of 1.5–2.5 µm and 6.47×10^{18} cm^{-3}.

Keywords: CMZTSSe films; sol–gel; electrical properties; optical properties; selenization treatment; solar cells

---

1. Introduction

Thin-film photovoltaic cells have generated enormous attention since the reliable efficiencies of $CuInGaSe_2$ (CIGSe) and CdTe have exceeded 20% [1–4]. While there are numerous benefits of CIGSe and CdTe photovoltaic cells, such as consuming less material and high efficiency, the constituent elements of the materials inevitably have high cost and toxicity. Compared with CIGSe and CdTe, the $Cu_2ZnSn(S,Se)_4$ (CZTSSe) compound has a high absorption coefficient and adjustable bandgaps, and the constituent elements are inexpensive and environmentally friendly [5,6]. Hence, CZTSSe is considered to be a potential absorber material. However, the conversion efficiency of CZTSSe photovoltaic cells hasonly achieved 12.6%, which is still far lower than the conversion efficiency of CIGSe-based solar cells (22.9%) [7–9]. In order to improve conversion efficiency and make CZTSSe industrially viable, a lot of researches are still needed. Recently, studies have shown that enhancing the open-circuit voltage (V_{oc}) and improving the crystal quality and optical–electrical characteristics of the CZTSSe layer are the main challenges that CZTSSe photovoltaic cells must face [10–12]. Considerable researches have been carried out to explore the reason of the lower V_{oc}. It was found that the reasons are varied, and one of the frequently mentioned ones is the unsuitable conduction band offset (CBO) [10–12]. The CBO of CdS/CZTSSe is affected by the bandgaps of CZTSSe. Therefore, engineering an adjustable bandgap is a practical method to break through the current limit of V_{oc}. In order to achieve the goal of tuning CZTSSe bandgaps, a large number of experiments have been carried out [13–18].

Recent studies show that adjusting the bandgaps of CZTSSe by partial metal cation replacement is probably an effective approach. Among them, taking the place of Zn with Mg can adapt the bandgaps and enhance the crystallinity of CZTSSe films. In our previous studies, the $Cu_2Mg_xZn_xSn(S,Se)_4$ samples with different Mg concentrations were successfully synthesized by the sol–gel method [19]. It was found that the bandgaps of $Cu_2Mg_xZn_xSn(S,Se)_4$ samples can be tuned in the ranges of 1.12 to 0.88 eV as Mg concentration varied from x = 0 to 0.6 [19]. In addition, we investigated the effects of Mg content on the properties of $Cu_2Mg_xZn_xSn(S,Se)_4$ films in detail. The results of the study showed that the $Cu_2Mg_xZn_xSn(S,Se)_4$ films with adjusted bandgaps, high crystallinity, and high carrier concentration will be a potential high-efficiency photovoltaic cell absorber material [19]. Furthermore, the realization of bandgap regulation by Mg instead of Zn has the following advantages: Firstly, compared to Cd and Ge, the Mg element is more abundant, low cost, and environmentally friendly [20]. Moreover, the formation of other binary and ternary phases may be eliminated or reduced in the process of synthesizing $Cu_2Mg_xZn_xSn(S,Se)_4$ films, because ZnS and ZnSe are present, while MgS and MgSe are unstable in the process of synthesizing the precursor solution [20]. Therefore, it is concluded that $Cu_2Mg_xZn_xSn(S,Se)_4$ is worthy of study as a potential absorbing layer material.

As we all know, for the sake of improving the properties of CZTSSe and obtaining CZTSSe films with a single phase and large crystal size, a heat treatment of the precursor films at an elevated temperature (>500 °C) is usually required [21,22]. A large number of studies have shown that the selenization temperature and selenization time significantly influence the properties of the films, including the crystal quality, and optical and electrical properties [23–25]. The influence of selenization treatment on the physical performance of CZTSSe has been investigated by extensive researches [21–25]. However, the effects of selenization temperature and selenization time on the phase evolutions, crystal quality, and optical–electrical characteristics of $Cu_2Mg_xZn_xSn(S,Se)_4$ have not been reported so far. Hence, in the present work, $Cu_2Mg_{0.2}Zn_{0.8}Sn(S,Se)_4$ (CMZTSSe) samples annealed at different selenization conditions are synthesized, and the influences of selenization temperature and selenization time on the structure, and optical and electrical properties of CMZTSSe samples are investigated in detail.

2. Experimental Methods

2.1. Synthesis of $Cu_2Mg_{0.2}Zn_{0.8}Sn(S,Se)_4$ Precursor Films

The CMZTSSe precursor films were synthesized in two steps. The first process was to prepare the CMZTSSe precursor solution by a simple and convenient sol–gel technique. We dissolved $Cu(CH_3COO)_2·H_2O$ (0.8086 g), $MgCl_2·6H_2O$ (0.1787 g), $Zn(CH_3COO)_2·2H_2O$ (0.4794 g), $SnCl_2·2H_2O$ (0.5077 g), and thiourea (1.3702 g) into 2-methoxyethanol (10 mL), and stirred for 10–15 min at room temperature. The monoethanolamine (MEA) (0.2 mL) was added to the precursor solutions at the end. During the precursor solution preparation process, in order to obtain high-quality CMZTSSe films, the ratio of Cu/(Mg + Zn + Sn) was 0.82 and the ratio of (Zn + Mg)/Sn was 1.2. In our previous studies, we found that the surface morphology, and optical and electrical properties of $Cu_2Mg_xZn_xSn(S,Se)_4$ films were optimal when the proportion of Mg/(Mg + Zn) was 0.2 [17]. Therefore, in the present work, the proportion of Mg/(Mg + Zn) was set to 0.2. Then, in order to dissolve the raw materials completely and obtain the CMZTSSe sol–gel solution at room temperature, we stirred the solution until the precursor solution color became colorless and transparent. The second procedure was to obtain the CMZTSSe precursor films through the spin coating method. We spun the precursor solution of CMZTSSe at 3000 r for 30 s, followed by drying at 300 °C for 5 min in air. In order to acquire CMZTSSe films with micrometer thicknesses, the process of the coating and drying was repeated several times.

2.2. Selenization of $Cu_2Mg_{0.2}Zn_{0.8}Sn(S,Se)_4$ Films

For the sake of studying the effect of annealing treatment on the properties of CMZTSSe films, the rapid annealing treatment was implemented for CMZTSSe precursor films at various selenization temperatures and selenization times under the same selenium atmosphere. We used fixed-quality selenium powder (15 mg) to anneal the precursor CMZTSSe films in the selenide annealing furnace, and increased the selenization temperature from 500 to 560 °C, while the selenization time was adjusted in the range of 10–20 min to obtain CMZTSSe samples at varied annealing conditions.

2.3. Materials Characterization

The structural characteristics and chemical composition of CMZTSSe were measured using X-ray power diffraction (XRD, Rigaku Corporation, Tokyo, Japan), Raman spectroscopy (Renishaw, London, UK) with a 514 nm laser wavelength, and X-ray photoelectron spectroscopy (XPS, Thermo Fisher Scientific, Waltham, MA, USA) (Al Kα was used as the X-ray source). Scanning electron microscopy (SEM) (Hitachi S-4800, JEOL Ltd., Tokyo, Japan) was performed to study the surface morphology of CMZTSSe films. The energy-dispersive X-ray spectroscopy (EDS, JEOL Ltd., Tokyo, Japan) system was used to analyze elemental content. The optical and electrical performances of CMZTSSe films were characterized by UV-Vis-NIR spectra (UV-3101PC, Tokyo, Japan) and room temperature Hall measurement, respectively.

3. Results and Discussion

3.1. Influence of Annealing Temperature on the Properties of $Cu_2Mg_{0.2}Zn_{0.8}Sn(S,Se)_4$ Films

As we all know, the properties of absorbers are easily affected by annealing temperature. For the sake of studying the effect of selenization temperature on the microstructure and photoelectric characteristics of CMZTSSe films during the selenization process, the CMZTSSe samples were annealed for 15 min under Se ambience at different temperatures of 500, 530, and 560 °C, hereafter named as A1, A2, and A3, respectively. In addition, the CZTSSe film annealed at a temperature of 530 °C for 15 min will be used as a reference, named as sample A.

XRD spectra were always used to evaluate the crystal quality and assess the probable impurity phase during the selenization process. Figure 1 represents the XRD spectra of CMZTSSe samples annealed at different temperatures from 500 to 560 °C (samples A1, A2, and A3). The CMZTSSe films at all varied annealing temperatures were in the kesterite phase, with peaks corresponding to the (112), (220), (312), (008), and (332) planes of Cu_2ZnSnS_4 (CZTS) with the kesterite phase [26,27]. As shown in Figure 1, the characteristic peaks of impurity phases were not found, which indicates that the single-phase CMZTSSe with kesterite structure was formed at all varied annealing temperatures. Figure 1a indicates the XRD patterns for sample A1, annealed at 500 °C under atmosphere of Se, and it was found that the peak intensity is weaker. By increasing the annealing temperature to 530 °C, as shown in Figure 1b, the peak intensity was found to be enhanced, indicating that the crystal quality of sample A2 is enhanced at 530 °C. For sample A3, with the annealing temperature of 560 °C, the peak intensity of XRD only slightly changed, as seen from Figure 1c. As seen from the inset, the dominant characteristic peak (112) is observed to be shifted to smaller angles, from 27.20° to 26.93°, with the annealing temperature increasing from 500 to 560 °C. This is due to the variety in atomic-lattice distance caused by elemental replacement, where the element Se will take the place of S in the CMZTSSe compound with the increase of annealing temperature. According to the results of XRD, there are no secondary phases in CMZTSSe films annealed at different selenization temperatures from 500 to 560 °C.

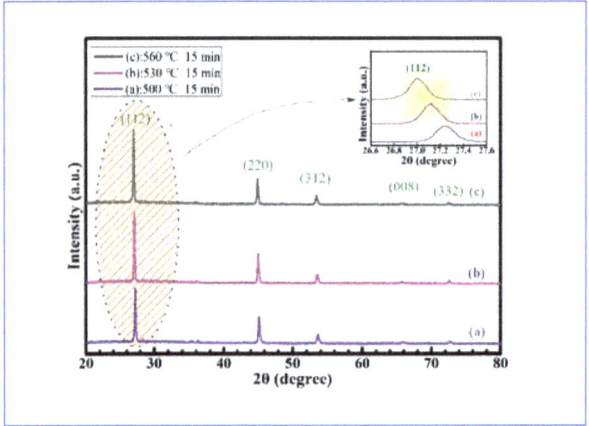

Figure 1. XRD spectra of $Cu_2Mg_{0.2}Zn_{0.8}Sn(S,Se)_4$ (CMZTSSe) films annealed at different temperatures (**a**) 500 °C, (**b**) 530 °C, (**c**) 560 °C. Inset: Enlarged view of the corresponding (112) diffraction peak of the CMZTSSe films annealed at different temperatures.

Table 1 shows the values of full width at half-maximum (FWHM), grain size, the a-axis lattice constant (a), and the c-axis lattice constant (c) for the CMZTSSe films annealed at different temperatures, which are obtained from the (112) peak in the XRD profile. It can be seen from Table 1 that, as the annealing temperature increases from 500 to 560 °C, the grain size increases first and attains a maximum value at 530 °C, and then decreases when the annealing temperature is continuously increased up to 560 °C. An opposite changing trend was observed concurrently for the FWHM—when increasing the annealing temperature up to 530 °C, the FWHM has a minimum value. Meanwhile, the a-axis lattice constant gradually increases from 5.669 to 5.691 Å as the annealing temperature increases from 500 to 560 °C. The c-axis lattice constant also increases from 11.305 to 11.635 Å. The increases in the lattice constants are ascribed to the rise of the Se content in films, where the S (0.184 nm) atoms were replaced by the larger Se (0.198 nm) atoms with the increasing annealing temperature. This may be verified by the EDS results later.

Table 1. The full width at half-maximum (FWHM), grain size, a-axis lattice constant (a), and c-axis lattice constant (c) for CMZTSSe films annealed at different temperatures.

Sample	Temperature (°C)	Time (min)	a (Å)	c (Å)	Crystalline Size (nm)	FWHM
A1	500	15	5.669	11.305	58.2	0.172
A2	530	15	5.681	11.415	71.1	0.156
A3	560	15	5.691	11.635	67.1	0.163

It is well known that the discernment of the secondary phases in CZTSSe compounds by XRD patterns is difficult, owing to the nearly overlapped XRD patterns of the secondary ZnS(Se) and $Cu_2SnS(Se)_3$ with the kesterite CZTSSe [28]. Therefore, Raman spectroscopy measurement is usually applied as an auxiliary technique to detect possible impurity phases, because it is sensitive to lattice vibrations.

Figure 2 shows the Raman spectra of CMZTSSe films annealed at different annealing temperatures from 500 to 560 °C (samples A1, A2, and A3). The Raman spectrum of sample A1 was fitted using the Gaussian fitting method, and the peaks at 173, 193, and 234 cm^{-1} can be observed. The peaks at 173 and 193 cm^{-1} conform to the A (A1 and A2) mode Raman vibration peaks of the kesterite CZTSSe phase, as reported in the previous literature [29]. The A modes are pure anion modes which correspond to vibrations of pure chalcogen (S or Se) atoms surrounded by motionless neighboring atoms. The result exhibits a broad peak located at 234 cm^{-1} that corresponds to the E vibration mode in connection with Sn–Se bonding in the kesterite CZTSSe phase [29]. In addition, it can be seen that the Raman peaks of all samples conform with the vibration peaks of the kesterite CZTSSe phase. The Raman peaks of the other secondary phases were not discovered. This indicates that all annealed samples consist of a single phase of CZTSSe with kesterite structure. Furthermore, with the variation of annealing temperature from 500 to 560 °C, it was found that all Raman peaks were shifted to the lower values, especially for the A1 mode peak. The inset of Figure 2 displays the tracking of the peak position of the A1 vibration mode, and it is obviously noted that the A1 vibration peak shifted from 195.6 to 193.9 cm^{-1} as the annealing temperature changed from 500 to 560 °C. This phenomenon can be attributed to the increment of selenization temperature from 500 to 560 °C, which easily allows larger Se to replace S in the CZTSSe, significantly increasing the lattice constant. It was found from the Raman results that the secondary phases were not discovered, which is consistent with the XRD results. In addition, the peak shift as observed from the XRD and Raman spectra is considered to be due to larger Se taking the site of smaller S in CMZTSSe films with the increase of selenization temperature from 500 to 560 °C, which will be better understood from the compositional and morphological studies later.

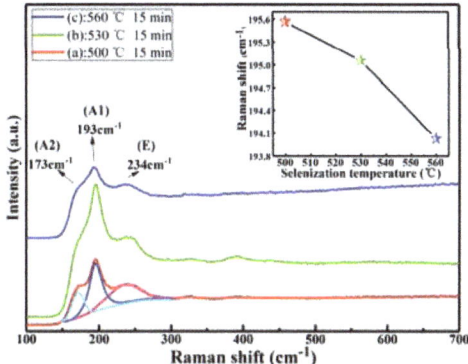

Figure 2. Raman spectra of the CMZTSSe films annealed at different temperatures. Inset: The main Raman peaks of A1 mode for CMZTSSe films annealed at different temperatures.

XPS is sensitive to information about the chemical bonding state and element content. As shown in Figure 3, we identify the elemental composition and valence states of the constituent elements (Cu, Zn, Sn, Se, S, and Mg) in CMZTSSe films annealed at 530 °C for 15 min by XPS measurement, and all peaks were corrected by the C1s binding energy (284.8 eV). As seen from Figure 3a, the Cu 2p XPS spectrum consists of a Cu 2p$_{3/2}$ peak and a Cu 2p$_{1/2}$ peak at 931.6 and 952.5 eV, respectively, with a peak separation of 20.9 eV, which indicates that Cu is in the state of +1 [30]. Figure 3b displays a Zn 2p XPS spectrum, and the peaks presented at 1021.4 and 1044.2 eV are ascribed to Zn 2p$_{3/2}$ and Zn 2p$_{1/2}$, respectively. The binding energy interval between the Zn 2p$_{3/2}$ peak and the Zn 2p$_{1/2}$ peak is 22.8 eV, indicating the existence of divalent Zn ions [31]. Figure 3c shows the XPS spectrum of Sn 3d, and two peaks attributed to Sn 3d$_{5/2}$ and Sn 3d$_{3/2}$ can be observed at 485.5 and 494.1 eV. The energy difference of the two Sn 3d peaks is 8.6 eV, suggesting the presence of the Sn^{4+} state [32]. Figure 3d displays the Se 3d XPS spectrum, and the S 2p high-resolution spectrum is shown in Figure 3e. The Se 3d XPS spectra can be fitted into two sub-peaks located at 53.3 and 53.8 eV, which can be ascribed to Se 3d$_{3/2}$ (green area) and Se 3d$_{1/2}$ (purple area), respectively. The above results indicate that Se in the films is likely to exist in the Se^{2-} state [33]. It is well known that the S 2p core level and Se 3p core level are almost overlapping, and we used the Gaussian fitting method to fit the XPS spectra into four sub-peaks presented at 159.1, 160.2, 161.3, and 165.8 eV, which are ascribed to Se 2p$_{3/2}$ (light green area), S 2p$_{3/2}$ (pink area), S 2p$_{1/2}$ (deep purple area), and Se 3p$_{1/2}$ (grey green area), respectively. The S 2p$_{3/2}$ and S 2p$_{1/2}$ peaks located at 160.2 and 161.3 eV, respectively, are in the standard reference value range (160–164 eV) [33], which means that S exists in the form of S^{2-}. Figure 3f displays the XPS high-resolution spectrum of Mg 1s, and the peak can be observed at 1303.6 eV, which indicates that divalent Mg^{2+} exists in our study [19]. The results of the XPS analysis show that the constituent elements (Cu, Zn, Sn, Se, S, and Mg) exist in the forms of Cu^{1+}, Zn^{2+}, Mg^{2+}, Sn^{4+}, Se^{2-}, and S^{2-} in CMZTSSe.

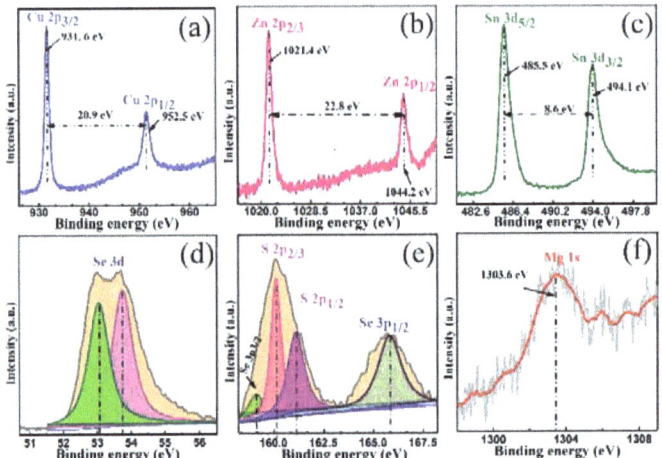

Figure 3. X-ray photoelectron spectroscopy (XPS) spectra of CMZTSSe films annealed at 530 °C for 15 min: (a) Cu, (b) Zn, (c) Sn, (d) Se, (e) S, and (f) Mg.

Figure 4a–d shows the scanning electron microscopy (SEM) images of the CZTSSe film annealed at 530 °C for 15 min (sample A) and the CMZTSSe films annealed under atmosphere of Se for 15 min at temperatures of 500, 530, and 560 °C (samples A1, A2, and A3). Figure 4a shows the surface SEM images of sample A. As we can see from Figure 4a, irregular and small grain sizes of 0.5–0.8 µm were observed. In addition, it was clearly seen that the surface morphology of $Cu_2ZnSn(S,Se)_4$ films is very rough. Figure 4b shows the SEM of sample A1, and it is clearly seen that the film consists of nanograins, with the grain size being even smaller than that of sample A, and the surface morphology is also rough. As shown in Figure 4c, the crystal quality was improved for sample A2, and the grain size of the film reached 1.0–2.5 µm while the surface morphology became smooth and compact. When the selenization temperature was increased to 560 °C, the grain size of sample A3 slightly reduced to 1.0–1.5 µm and displayed a rough morphology, as displayed in Figure 4d. The results indicate that the proper selenization temperature is 530 °C, and that this is beneficial to promote the grain growth. Excessively high or low temperatures will cause the deterioration of film quality. When the selenization temperature is 530 °C, not only does the grain size of CMZTSSe film reach its maximum, but the surface morphology of the film also becomes smooth and compact.

As we all know, the photoelectric properties of CZTSSe-based films need to rely heavily on the stoichiometric ratios of Cu, Zn, Sn, S, and Se in CZTSSe-based films [34]. Table 2 displays the EDS results of CMZTSSe films annealed at different temperatures from 500 to 560 °C (samples A1, A2, and A3). According to the EDS results, we confirmed the existence of Cu, Zn, Mg, Sn, S, and Se elements in samples A1, A2, and A3. It was found that the atomic percentages of Se increased from 35.87% to 45.20% and S evidently decreased from 11.46% to 3.16% with increasing selenization temperature from 500 to 560 °C, indicating that Se will partially replace S in the CMZTSSe compound. The compositions of the other elements (Cu, Sn, and Mg) in all three samples were found to be nearly the same, while the atomic percentages of Zn decreased from 11.30% to 8.16%, indicating that Zn loss happened when the selenization temperature changed from 500 to 560 °C. In the precursors, the ratios of Cu/(Zn + Mg + Sn) and Mg/(Mg + Zn) were about 0.82 and 0.2, respectively. The ratios of Mg/(Mg + Zn) in all the films were close to 0.2, but the ratios of Cu/(Zn + Mg + Sn) in all films became significantly larger. This may be due to the decrease of Zn content as the annealing temperature increases. As mentioned before, Se/(S + Se) > 50% is highly suitable for the fabrication of high-efficiency solar cells [35]. The percentage of Se/(S + Se) is 75.79% for A1, and the percentages of Se/(S + Se) increase to 82.22% and 93.47% for A2 and A3, respectively, indicating that A1, A2, and A3 samples are suitable for the fabrication of efficient solar cell devices. Sample A2 has an appropriate Se/(S + Se) ratio and the grain size became larger compared to samples A1 and A3. Therefore, sample A2 is more suitable to fabricate the high-efficiency solar cells. Figure 5 summarizes the elements composition analysis of CMZTSSe films according to Table 2. As shown from Figure 5, when increasing the selenization temperature from 500 to 560 °C, the proportion of Se increases gradually while the content of S decreases, while the atomic percentages of Cu, Zn, Sn, and Mg remain relatively constant compared with those of S and Se.

Figure 4. SEM images of CZTSSe annealed at (**a**) 530 °C and CMZTSSe films annealed at (**b**) 500; (**c**) 530; and (**d**) 560 °C.

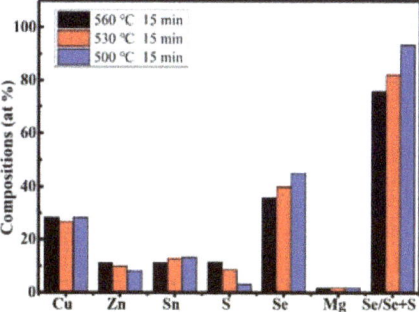

Figure 5. Energy-dispersive X-ray spectroscopy (EDS) composition analyses of CMZTSSe films annealed at different temperatures.

Table 2. EDS results of the CMZTSSe films annealed at different temperatures from 500 to 560 °C.

Sample	Temperature (°C)	Cu (at%)	Zn (at%)	Sn (at%)	Mg (at%)	S (at%)	Se (at%)	Se/(S + Se)	Cu/(Zn + Mg + Sn)	Mg/(Mg + Zn)
A1	500	28.46	11.30	11.14	1.78	11.46	35.87	75.79	1.18	0.14
A2	530	26.71	9.91	12.88	1.76	8.67	40.08	82.22	1.09	0.15
A3	560	28.32	8.16	13.28	1.88	3.16	45.20	93.47	1.22	0.19

In order to research the influence of selenization temperature on the optical bandgaps of CMZTSSe films (samples A1, A2, and A3), we studied the optical absorption measurements of the CMZTSSe films annealed at different selenization temperatures by an UV-vis-NIR spectrophotometer. Figure 6 displays the $(\alpha h\upsilon)^2$–$h\upsilon$ plots of CMZTSSe films. We use the solids band theory to express the relation between the absorption coefficient (α) and the photon energy ($h\upsilon$) as follows [36]:

$$(\alpha h\upsilon) = B(h\upsilon - E_g)^n \qquad (1)$$

where h, B, υ, and E_g are Plank's constant, a constant, photon frequency, and optical bandgap, respectively. The values of n can employ 3, 2, 3/2, and 1/2, when transitions are indirect unallowed, indirect allowed, direct unallowed, and direct allowed, respectively [37]. The values of n can employ 1/2 for direct bandgaps of semiconductor CZTSSe [36]. By using the Equation (1) and the data in Figure 6, the bandgaps of CMZTSSe are evaluated to be 1.04, 1.02, and 0.93 eV for samples A1, A2, and A3, respectively, as shown in the illustration of Figure 6. It was found that the bandgap values of samples A1, A2, and A3 gradually decrease with increasing selenization temperature, which can be attributed to the changes of crystal lattice and disparities in electronegativities owing to alloying and modified atomic structures through Se taking the site of S.

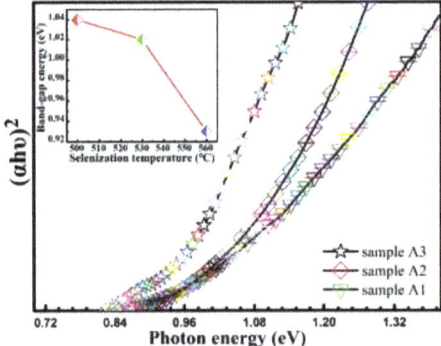

Figure 6. The plot of $(\alpha h\upsilon)^2$ vs. $h\upsilon$ for the absorption spectra. Inset: Bandgap variation as a function of the selenization temperature.

The electrical properties of the absorbing layer are also important factors affecting the efficiency of solar cells. Table 3 displays the electrical characteristics of CMZTSSe annealed at different temperatures (samples A1, A2, and A3) by the Vander Paw method at room temperature. It was found that the CMZTSSe films annealed at different temperatures behave with p-type semiconductor characteristics. When the selenization temperature is increased from 500 to 530 °C, the resistivity first decreases from 6.18×10^0 to 2.85×10^{-1} $\Omega \cdot$cm, then increases to 1.10×10^2 $\Omega \cdot$cm at the selenization temperature of 560 °C. Obviously, when the selenization temperature is 530 °C, the resistivity is optimal. Simultaneously, it is clear that the corresponding carrier concentration shows a best value of 6.47×10^{18} cm^{-3} at the selenization temperature of 530 °C. In addition, the mobility reduced from 1.09×10^0 cm^{2}V^{-1}s^{-1} (500 °C) to 3.31×10^{-1} cm^{2}V^{-1}s^{-1} (530 °C), but increased to 1.04×10^{-1} cm^{2}V^{-1}s^{-1} at the selenization temperature of 560 °C. We analyzed the reasons for the change of CMZTSSe electrical properties, combined with the characterization of SEM. It was concluded that the defects at the surfaces of the absorbing layers are passivated, and owing to the crystal quality of CMZTSSe films improving with the selenization temperature increasing from 500 to 530 °C, the resistivity and carrier concentration achieve the best values with sample A2. It is obvious that the deterioration of resistivity and carrier concentration is owing to the deterioration of crystal quality when the selenization temperature further increases from 530 to 560 °C.

Table 3. Electrical properties of the CMZTSSe films annealed at different temperatures from 500 to 560 °C.

Sample	Temperature (°C)	Time (min)	ρ (Ω·cm)	n (cm^{-3})	μ (cm^{-2}V^{-1}s^{-1})	Conduction Type
A1	500	15	6.18×10^0	9.22×10^{17}	1.09×10^0	p
A2	530	15	2.85×10^{-1}	6.47×10^{18}	3.31×10^{-1}	p
A3	560	15	1.10×10^2	5.54×10^{17}	1.04×10^{-1}	p

3.2. Effect of Selenization Time on Properties of CMZTSSe Films

After a series of analyses and characterizations, it was proved that the best selenization temperature is 530 °C. As we all know, selenization time is also one of the major parameters influencing the performance of the absorber layer. In our study, after optimizing the annealing temperature, the influence of annealing time on the properties of CMZTSSe films has been studied. The CMZTSSe films were annealed for 10, 15, and 20 min at 530 °C under atmosphere of Se, afterward referred as B1, B2, and B3, respectively.

Figure 7a–c shows the XRD patterns of CMZTSSe films annealed for different times from 10 to 20 min (samples B1, B2, and B3). For the samples B1, B2, and B3, five diffraction peaks located at 28.53°, 47.33°, 56.17°, 69.27°, and 76.44° were observed, conforming to the (112), (220), (312), (008), and (332) planes of kesterite CZTS respectively [25,26]. The characteristic diffraction peaks of other impurity phases were not observed in Figure 7. We can explore the influence of annealing time on the structural performance of CMZTSSe films by observing the peak intensity and peak shift. It is observed from Figure 7a,b that the intensity of the (112) peak is increased, which indicates that the crystal quality is enhanced by increasing the annealing time from 10 to 15 min. Furthermore, the characteristic peak intensity is almost unchanged with the increase of annealing time from 15 to 20 min, as displayed in Figure 7c. In addition, the position of the (112) peaks are shifted to lower 2θ angle with selenization time increasing, as shown in the illustration of Figure 7. Since the concentration of Se in the CMZTSSe matrix increases with the increase of selenization time, there is enlargement of unit cell size, causing the change of the lattice distance in the films. According to the analysis results of XRD, it was found that the crystal structure of CMZTSSe was not changed with the increase of annealing time, and we speculate that the crystal growth of CMZTSSe was completed when the annealing time reached 15 min.

The full width at half-maximum (FWHM), grain size, a-axis lattice constant (a), and c-axis lattice constant (c) of the films annealed for different times from 10 to 20 min (samples B1, B2, and B3) are displayed in Table 4. It was found that by increasing the annealing time from 10 to 20 min, the grain size first increases and then decreases, while the FWHM value first decreases and then increases. When the annealing time is 15 min, the grain size has a maximum value, while the FWHM has a minimum value. The a-axis lattice constant gradually increases from 5.665 to 5.684 Å as the annealing time increases from 10 to 20 min. Meanwhile, the c-axis lattice constant also increases from 11.334 to 11.453 Å. It was concluded that the optimal annealing time is 15 min.

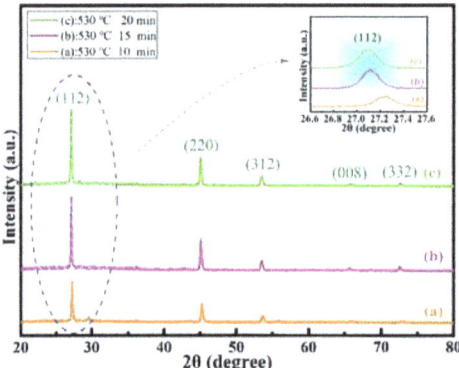

Figure 7. XRD spectra of CMZTSSe films annealed at 530 °C for different time (**a**) 10 min, (**b**) 15 min, (**c**) 20 min. Inset: Enlarged view of the corresponding (112) diffraction peaks of the CMZTSSe films annealed for different times.

Table 4. The full width at half-maximum (FWHM), grain size, a-axis lattice constant (a), and c-axis lattice constant (c) for CMZTSSe films annealed for different times.

Sample	Temperature (°C)	Time (min)	a (Å)	c (Å)	Crystalline Size (nm)	FWHM
B1	530	10	5.665	11.334	52.8	0.190
B2	530	15	5.681	11.415	71.1	0.156
B3	530	20	5.684	11.453	61.8	0.171

As we all know, the diffraction peaks of Cu_3SnS_4 with a tetragonal structure and ZnS with a cubic structure are close to the diffraction peaks of CZTSSe with a kesterite structure [38,39]. Thus, it is very difficult to detect possible secondary phases by XRD only. In order to further detect the phase compositions of CMZTSSe films, Raman scattering spectra were measured for the films annealed for different times from 10 to 20 min (samples B1, B2, and B3), as shown in Figure 8. The Raman spectrum of sample B1 was fitted using the Gaussian fitting method, and three Raman peaks located at ~193 (A1 mode), 173 (A2 mode), and 234 cm^{-1} (E mode) were observed, which conform to the Raman characteristic peaks of CZTSSe [29]. The characteristic peaks of some possible impurity phases were not detected. It is suggested that the CMZTSSe films are composed of a single phase of kesterite CZTSSe. In addition, these characteristic peaks slightly shift to lower values as the selenization time increases, owing to the incorporation of Se in the CMZTSSe compound. It should be noted that the Raman spectra are consistent with the XRD patterns, and some possible secondary phases (Cu_2SnSe_3, $SnSe_2$, SnSe, ZnSe, and MgSe) were observed. Furthermore, the phase change of the CMZTSSe films did not occur with increasing selenization time from 10 to 20 min. It was concluded that the pure-phase CMZTSSe films were successfully synthesized.

Figure 9a–d depicts the SEM surface images of CZTSSe annealed at 530 °C for 15 min (sample A) and CMZTSSe films annealed at 530 °C for different times from 10 to 20 min (samples B1, B2, and B3), respectively. As shown in Figure 9a, the surface of the sample A displays pin-hole free morphology and small grain size between 500–800 nm, with a rough surface. Figure 9b shows the surface morphology of sample B1, and it was clearly observed that the CMZTSSe film still consists of nanograins, where the grain size is between 500 and 1000 nm with a large number of holes on the surface. As the annealing time increased to 15 min, obvious morphological change is observed (Figure 9c), with the grains size of sample B2 increasing sharply to the micron level (1.0–2.5 µm) and the surface becoming dense and flat. By further extending the annealing time to 20 min, the surface morphology of sample B3 becomes rough but still dense, and the grains size slightly decreases to 0.8–1.3 µm. It was concluded that the optimal selenization time is 15 min, and the CMZTSSe films obtained at this time have the best crystallinity, the largest crystal grain size, and their surfaces are dense and flat.

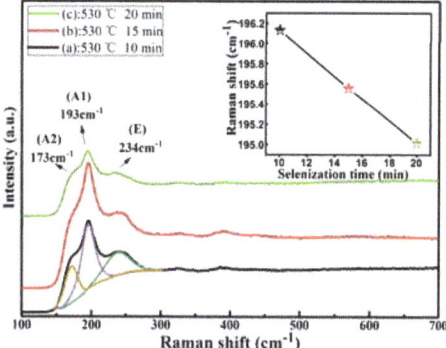

Figure 8. Raman spectra of the CMZTSSe films annealed for different times. Inset: The main Raman peaks of A1 mode for CMZTSSe films annealed for different times.

Figure 9. SEM images of CZTSSe annealed at 530 °C for (**a**) 15 min and CMZTSSe films annealed at 530 °C for (**b**) 10 min; (**c**) 15 min; and (**d**) 20 min.

The EDS results of the films annealed for different times from 10 to 20 min (samples B1, B2, and B3) are displayed in Table 5. It was found that by increasing the annealing time from 10 to 20 min, the atomic percentage of S evidently decreases from 10.23% to 2.26%, the atomic percentage of Se increases from 36.24% to 46.44%, and the ratio of Se/(Se + S) significant increases from 77.99% to 95.36%. The ratios of Cu/(Zn + Mg + Sn) in all films were significantly larger than those in the precursor, and the ratios of Mg/(Mg + Zn) in all films significantly decreased from 0.19 to 0.12 with increasing annealing time from 10 to 20 min. These changes were ascribed the decrease of Mg content with the increase of the annealing time, as shown in Table 5. Figure 10 represents the elemental composition analysis of the films annealed for different times from 10 to 20 min. It can be clearly seen that the atomic percentages of Se increase, while the atomic percentages of S decrease with the increase of annealing time. Compared with the increase of Se content and the decrease of S content, the atomic percentages of other elements (Cu, Zn, and Sn) changed only slightly. Furthermore, the changes were irregular and almost negligible, and the changes had less effect on the crystal quality of CMZTSSe films. According to the analysis of elemental composition, the change of selenization time mainly affects the atomic percentages of Se and S elements, and has little effect on other elements.

Table 5. EDS results of the CMZTSSe films annealed for different times from 10 to 20 min.

Sample	Time (min)	Cu (at%)	Zn (at%)	Sn (at%)	Mg (at%)	S (at%)	Se (at%)	Se/(S + Se)	Cu/(Zn + Mg + Sn)	Mg/(Mg + Zn)
B1	10	29.66	9.88	11.60	2.39	10.23	36.24	77.99	1.24	0.19
B2	15	26.71	9.91	12.88	1.76	8.67	40.08	82.22	1.09	0.18
B3	20	29.81	8.49	11.43	1.18	2.26	46.44	95.36	1.41	0.12

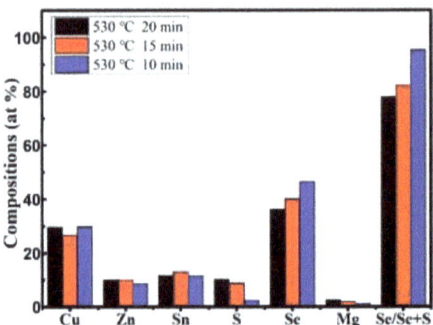

Figure 10. EDS composition analyses of CMZTSSe films annealed for different times.

Microstructure, composition, and grain size have great influences on the crystal quality of CMZTSSe, and the bandgaps are also crucial. According to the previous analysis, the change in the ratio of Se/(Se + S) with the increase of selenization time significantly influences the crystal quality, composition, and grain size. The effect of annealing time on the optical bandgaps of the CMZTSSe films has been evaluated by a UV-vis-NIR spectrophotometer. The theoretical basis of bandgap calculation is consistent with Formula 1. As shown in Figure 11, the bandgaps of the CMZTSSe films show a declining trend (1.06–0.95 eV) with increasing annealing time from 10 to 20 min. The illustration of Figure 11 displays the dependence of bandgaps on selenization time for CMZTSSe films. We can clearly see that the values of bandgap are 1.06, 1.02, and 0.95 for sample B1, sample B2, and sample B3, respectively. Combined with the analysis results of XRD, Raman, and EDS, the decline in bandgaps with the increase of the selenization time is ascribed to the increase of elemental Se.

As shown in Table 6, the impacts of selenization time on the conductivity, carrier concentration, and mobility of CMZTSSe (samples B1, B2, and B3) were investigated by Hall measurements at room temperature. It was observed that the p-type conductivity of CMZTSSe was not changed with the increase of annealing time from 10 to 20 min. As shown in Table 6, when the annealing time increased from 10 to 15 min, the hole concentration of the CMZTSSe films increased obviously from 9.22×10^{17} to 6.47×10^{18} cm^{-3}, the resistivity decreased from 6.18×10^{0} to 2.85×10^{-1} Ω·cm, and the mobility decreased from 1.09×10^{0} to 3.31×10^{-1} cm^{2}V^{-1}s^{-1}. Combined with the analysis results of SEM, by increasing annealing time from 10 to 15 min, the grain size of CMZTSSe becomes bigger and the surface becomes smooth and hole-free, which leads to the improvement of electrical performance. When the annealing time increases from 15 to 20 min, the crystallinity of CMZTSSe films deteriorates, and hence the hole concentration and resistivity decrease to 5.54×10^{17} cm^{-3} and 1.10×10^{2} Ω·cm, respectively. It was found that when the selenization temperature and selenization time are 530 °C and 10 min, the best electrical properties of the films are obtained.

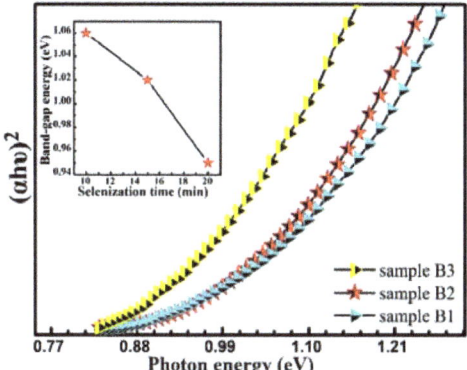

Figure 11. The plot of $(\alpha h\nu)^2$ vs. $h\nu$ for the absorption spectra. Inset: Bandgap variation as a function of the selenization time.

Table 6. Electrical properties of the CMZTSSe films annealed for different times from 10 to 20 min.

Sample	Temperature (°C)	Time (min)	ρ (Ω·cm)	n (cm^{-3})	μ (cm^{-2}V^{-1}s^{-1})	Conduction Type
B1	530	10	6.18×10^0	9.22×10^{17}	1.09×10^0	p
B2	530	15	2.85×10^{-1}	6.47×10^{18}	3.31×10^{-1}	p
B3	530	20	1.10×10^2	5.54×10^{17}	1.04×10^{-1}	p

4. Conclusions

In summary, we have successfully fabricated pure-phase CMZTSSe films at different selenization temperatures and selenization times through the sol–gel method. It was found that the properties of CMZTSSe films are greatly affected by selenization temperature and selenization time. Combined with the results of XRD, Raman, XPS, and EDS, it is clear that single-phase CMZTSSe films have been synthesized at different selenization temperatures and times, and the content of Se increases gradually while the content of S decreases gradually with increasing selenization temperature and selenization time. The SEM results suggested that the crystal quality of CMZTSSe is the best at the optimal selenization condition of 530 °C for 15 min, where the grain size reaches 1.0–2.5 μm. In addition, the grain-boundary passivation due to the crystal quality improvement will result in the improvement of electrical performance. The CMZTSSe films with p-type conductivity and high hole concentration of 6.47×10^{18} cm^{-3} were obtained by selenization at 530 °C for 15 min. The E_g of CMZTSSe films is decreased from 1.04 to 0.93 eV with increasing selenization temperature from 500 to 560 °C. When selenization time is increased from 10 to 20 min, the E_g of CMZTSSe can be adjusted from 1.06 to 0.96 eV. It is concluded that the structure, and optical and electrical properties of CMZTSSe will be optimal at an optimized selenization temperature and selenization time of 530 °C and 15 min, respectively, which will create an ideal absorber material for preparing higher efficiency kesterite solar cells.

Author Contributions: Conceptualization, Y.S. and Y.Z.; Writing-Original Draft Preparation, D.J.; Software, W.H.; Formal analysis, Y.S., Z.W. and L.Y.; Investigation, Y.S. and F.W.; Writing-Review & Editing, Y.S. and B.Y.

Funding: This research was funded by the National Natural Science Foundation of China under Grant No. 61,505,067, 61,605,059, 61,475,063, 61,775,081, 61,705,079.

Conflicts of Interest: The authors declare no conflict of interest.

References

1. Lokhande, A.C.; Chalapathy, R.B.V.; He, M.; Jo, E.; Gang, M.; Pawar, S.A.; Lokhande, C.D.; Kim, J.H. Development of Cu_2SnS_3 (CTS) thin film solar cells by physical techniques: A status review. *Sol. Energy Mater. Sol. Cells* **2016**, *153*, 84–107. [CrossRef]
2. Jackson, P.; Hariskos, D.; Wuerz, R.; Kiowski, O.; Bauer, A.; Friedlmeier, T.M.; Powalla, M. Properties of $Cu(In,Ga)Se_2$ solar cells with new record efficiencies up to 21.7%. *Phys. Status Solidi* **2015**, *9*, 28–31.
3. Park, J.Y.; Chalapathy, R.B.V.; Lokhande, A.C.; Hong, C.W.; Kim, J.H. Fabrication of earth abundant $Cu_2ZnSnSSe_4$ (CZTSSe) thin film solar cells with cadmium free zinc sulfide (ZnS) buffffer layers. *J. Alloy. Compd.* **2017**, *695*, 2652–2660. [CrossRef]
4. Nguyen, M.; Ernits, K.; Tai, K.F.; Ng, C.F.; Pramana, S.S.; Sasangka, W.A.; Batabyal, S.K.; Holopainen, T.; Meissner, D.; Neisser, A.; et al. ZnS buffffer layer for $Cu_2ZnSn(SSe)_4$ monograin layer solar cell. *Sol. Energy.* **2015**, *111*, 344–349. [CrossRef]
5. Ito, K.; Nakazawa, T. Electrical and Optical Properties of Stannite-Type Quaternary Semiconductor Thin Films. *Jpn. J. Appl. Phys.* **1998**, *27*, 2094–2097. [CrossRef]
6. Ramasamy, K.; Malik, M.A.; O'Brien, P. Routes to Copper Zinc Tin Sulfide Cu_2ZnSnS_4 a Potential Material for Solar Cells. *Chem. Commun.* **2012**, *48*, 5703–5714. [CrossRef] [PubMed]
7. Wang, W.; Winkler, M.T.; Gunawan, O.; Gokmen, T.; Todorov, T.K.; Zhu, Y.; Mitzi, D.B. Device characteristics of CZTSSe thin-Film solar cells with 12.6% efficiency. *Adv. Energy Mater.* **2014**, *4*, 1301465. [CrossRef]
8. Kato, T.; Wu, J.L.; Hirai, Y.; Sugimoto, H.; Bermudez, V. Record Efficiency for Thin Film Polycrystalline Solar Cells Up to 22.9% Achieved by Cs-Treated $Cu(In,Ga)(Se,S)_2$. *IEEE J. Photovolt.* **2018**, *9*, 325–330. [CrossRef]
9. Sui, Y.R.; Wu, Y.J.; Zhang, Y.; Wang, F.Y.; Gao, Y.B.; Lv, S.Q.; Wang, Z.W.; Sun, Y.F.; Wei, M.B.; Yao, B.; et al. Synthesis of simple, low cost and benign sol–gel $Cu_2In_xZn_{1-x}SnS_4$ alloy thin films: Influence of different rapid thermal annealing conditions and their photovoltaic solar cells. *RSC Adv.* **2018**, *8*, 9038–9048. [CrossRef]
10. Jiang, Y.H.; Yao, B.; Li, Y.F.; Ding, Z.H.; Xiao, Z.Y.; Yang, G.; Liu, R.J.; Liu, K.S.; Sun, Y.M. Effect of Cd content and sulfurization on structures and properties of Cd doped Cu_2SnS_3 thin films. *J. Alloy. Compd.* **2017**, *721*, 92–99. [CrossRef]
11. Kaur, K.; Kumar, N.; Kumar, M. Strategic review of interface carrier recombination in earth abundant Cu–Zn–Sn–S–Se solar cells: Current challenges and future prospects. *J. Mater. Chem. A* **2017**, *5*, 3069–3090. [CrossRef]
12. Rey, G.; Redinger, A.; Sendler, J.; Weiss, T.P.; Thevenin, M.; Guennou, M.; El Adib, B.; Siebentritt, S. The bandgap of $Cu_2ZnSnSe_4$: Effect of order-disorder. *Appl. Phys. Lett.* **2014**, *105*, 112106. [CrossRef]
13. Johnson, M.; Baryshev, S.V.; Thimsen, E.; Manno, M.; Zhang, X.; Veryovkin, I.V.; Leighton, C.; Aydil, E.S. AlkaliMetal-Enhanced Grain Growth in Cu_2ZnSnS_4 Thin Films. *Energy Environ. Sci.* **2014**, *7*, 1931–1938. [CrossRef]
14. Yuan, M.; Mitzi, D.B.; Gunawan, O.; Kellock, A.J.; Chey, S.J.; Deline, V.R. Antimony Assisted Low-Temperature Processing of $CuIn_{1-x}Ga_xSe_{2-y}S_y$ Solar Cells. *Thin Solid Films* **2010**, *519*, 852–856. [CrossRef]
15. Hsieh, Y.T.; Han, Y.T.; Jiang, C.; Song, T.B.; Chen, H.; Meng, L.; Zhou, H.; Yang, Y. Efficiency Enhancement of $Cu_2ZnSn(S,Se)_4$ Solar Cells via Alkali Metals Doping. *Adv. Energy Mater.* **2016**, *6*, 1502386. [CrossRef]
16. Tai, K.F.; Fu, D.C.; Chiam, S.Y.; Huan, C.H.A.; Batabyal, S.K.; Wong, L.H. Antimony Doping in Solution-processed $Cu_2ZnSn(S,Se)_4$ Solar Cells. *Chemsus Chem.* **2015**, *8*, 3504–3511. [CrossRef] [PubMed]
17. Khadka, D.B.; Kim, J.H. Structural Transition and Bandgap Tuning of $Cu_2(Zn,Fe)SnS_4$ Chalcogenide for Photovoltaic Application. *J. Phys. Chem. C* **2014**, *118*, 14227–14237. [CrossRef]
18. Khadka, D.B.; Kim, J.H. Structural optical and electrical properties of Cu_2FeSnX_4 (X=S, Se) thin films prepared by chemical spray pyrolysis. *J. Alloy. Compd.* **2015**, *638*, 103–108. [CrossRef]
19. Zhang, Y.; Sui, Y.R.; Wu, Y.J.; Jiang, D.Y.; Wang, Z.W.; Wang, F.Y.; Lv, S.Q.; Yao, B.; Yang, L.L. Synthesis and Investigation of environmental protection and Earth-abundant Kesterite $Cu_2Mg_xZn_{1-x}Sn(S,Se)_4$ thin films for Solar Cells. *Ceram. Int.* **2018**, *44*, 15249–15255. [CrossRef]
20. Wei, M.; Du, Q.; Wang, R.; Jiang, G.; Liu, W.; Zhu, C. Synthesis of New Earth-abundant Kesterite Cu_2MgSnS_4 Nanoparticles by Hot-injection Method. *Chem. Lett.* **2014**, *43*, 1149–1151. [CrossRef]
21. Xiao, Z.Y.; Yao, B.; Li, Y.F.; Ding, Z.H.; Gao, Z.M.; Zhao, H.F.; Zhang, L.G.; Zhang, Z.Z.; Sui, Y.R.; Wang, G. Influencing Mechanism of the Selenization Temperature and Time on the Power Conversion Efficiency of $Cu_2ZnSn(S,Se)_4$-based Solar Cells. *ACS Appl. Mater. Interfaces* **2016**, *8*, 17334–17342. [CrossRef] [PubMed]

22. Scragg, J.J.; Ericson, T.; Kubart, T.; Edoff, M.; Platzer-Björkman, C. Chemical Insights into the Instability of Cu$_2$ZnSnS$_4$ Films during Annealing. *Chem. Mater.* **2011**, *23*, 4625–4633. [CrossRef]
23. Hsu, C.J.; Duan, H.S.; Yang, W.; Zhou, H.; Yang, Y. Benign Solutions and Innovative Sequential Annealing Processes for High Performance Cu$_2$ZnSn(S,Se)$_4$ Photovoltaics. *Adv. Energy Mater.* **2014**, *4*, 1301287. [CrossRef]
24. Nguyen, D.C.; Ito, S.; Dung, D.V.A. Effects of Annealing Conditions on Crystallization of the CZTS Absorber and Photovoltaic Properties of Cu(Zn,Sn)(S,Se)$_2$ Solar Cells. *J. Alloy. Compd.* **2015**, *632*, 676–680. [CrossRef]
25. Ranjbar, S.; Rajesh Menon, M.R.; Fernandes, P.A. Effect of Selenization Conditions on the Growth and Properties of Cu$_2$ZnSn(S,Se)$_4$ Thin Films. *Thin Solid Films* **2015**, *582*, 188–192. [CrossRef]
26. Wu, Y.J.; Zhang, Y.; Sui, Y.R.; Wang, Z.W.; Lv, S.Q.; Wei, M.B.; Sun, Y.F.; Yao, B.; Liu, X.Y.; Yang, L.L. Bandgap engineering of Cu$_2$In$_x$Zn$_{1-x}$Sn(S,Se)$_4$ alloy films for photovoltaic applications. *Ceram. Int.* **2018**, *44*, 1942–1950. [CrossRef]
27. Kishor Kumar, Y.B.; Suresh Babu, G.; Uday Bhaskar, P.; Sundar Raja, V. Preparation and characterization of spray-deposited Cu$_2$ZnSnS$_4$ thin films. *Sol. Energy Mater. Sol. Cells* **2009**, *93*, 1230–1237. [CrossRef]
28. Walsh, A.; Chen, S.; Wei, S.H.; Gong, X.G. Kesterite Thin-Film Solar Cells: Advances in Materials Modelling of Cu2ZnSnS4. *Adv. Energy Mater.* **2012**, *2*, 400–409. [CrossRef]
29. Khadka, D.B.; Kim, J.H. Bandgap Engineering of Alloyed Cu$_2$ZnGe$_x$Sn$_{1-x}$Q$_4$ (Q=S,Se) Films for Solar Cell. *J. Phys. Chem. C* **2015**, *119*, 1706–1713. [CrossRef]
30. Rondiya, S.; Wadnerkar, N.; Jadhav, Y.; Jadkar, S.; Haram, S.; Kabir, M. Structural, Electronic, and Optical Properties of Cu$_2$NiSnS$_4$: A Combined Experimental and Theoretical Study toward Photovoltaic Applications. *Chem. Mater.* **2017**, *29*, 3133–3142. [CrossRef]
31. Calderón, C.; Gordillo, G.; Becerra, R.; Bartolo-Pérez, P. XPS analysis and characterization of thin films Cu$_2$ZnSnS$_4$ grown using a novel solution based route. *Mater. Sci. Semicond. Process.* **2015**, *39*, 492–498. [CrossRef]
32. Tsega, M.; Dejene, F.B.; Kuo, D.H. Morphological evolution and structural properties of Cu$_2$ZnSn(S,Se)$_4$ thin films deposited from single ceramic target by a one-step sputtering process and selenization without H$_2$Se. *J. Alloy. Compd.* **2015**, *642*, 140–147. [CrossRef]
33. Liu, W.C.; Guo, B.L.; Wu, X.S.; Zhang, F.M.; Mak, C.L.; Wong, K.H. Facile hydrothermal synthesis of hydrotropic Cu$_2$ZnSnS$_4$ nanocrystal quantum dots: Band-gap engineering and phonon confinement effect. *J. Mater. Chem. A* **2013**, *1*, 3182–3186. [CrossRef]
34. Xiao, Z.Y.; Li, Y.F.; Yao, B.; Ding, Z.H.; Deng, R.; Zhao, H.F.; Zhang, L.G.; Zhang, Z.Z. Significantly Enhancing the Stability of a Cu$_2$ZnSnS$_4$ Aqueous/Ethanol-based Precursor Solution and Its Application In Cu$_2$ZnSn(S,Se)$_4$ Solar Cells. *RSC Adv.* **2015**, *5*, 103451–103457. [CrossRef]
35. Shannon, R.; Parkinson, B.A.; Prieto, A.L. Compositionally Tunable Cu$_2$ZnSn(S$_{1-x}$Se$_x$)$_4$ Nanocrystals: Probing the Effect of Se-Inclusion in Mixed Chalcogenide Thin Films. *J. Am. Chem. Soc.* **2011**, *133*, 15272–15275.
36. Pankove, I.V. *Optical Processes in Semiconductors*; Dover Inc.: New York, NY, USA, 1975; pp. 34–95.
37. Chen, S.; Walsh, A.; Yang, J.H.; Gong, X.G.; Sun, L.; Yang, P.X.; Chu, J.H.; Wei, S.H. Compositional dependence of structural and electronic properties of Cu$_2$ZnSn(S,Se)$_4$ alloys for thin film solar cells. *Phys. Rev. B* **2011**, *83*, 1252011–1252015.
38. Fernandes, P.A.; Salomé, P.M.P.; Cunha, A.F.D. A Study of Ternary Cu$_2$SnS$_3$ and Cu$_3$SnS$_4$ Thin Films Prepared by Sulfurizing Stacked Metal Precursors. *J. Phys. D Appl. Phys.* **2010**, *43*, 215403. [CrossRef]
39. Riha, S.C.; Parkinson, B.A.; Prieto, B.A. Solution-Based Synthesis and Characterization of Cu$_2$ZnSnS$_4$ Nanocrystals. *J. Am. Chem. Soc.* **2009**, *131*, 12054–12055. [CrossRef]

© 2019 by the authors. Licensee MDPI, Basel, Switzerland. This article is an open access article distributed under the terms and conditions of the Creative Commons Attribution (CC BY) license (http://creativecommons.org/licenses/by/4.0/).

Review

Colloidal Self-Assembly of Inorganic Nanocrystals into Superlattice Thin-Films and Multiscale Nanostructures

Hongseok Yun [1],* and Taejong Paik [2],*

[1] Department of Chemical and Biomolecular Engineering, Korea Advanced Institute of Science and Technology (KAIST), Daejeon 34141, Korea
[2] Department of Integrative Engineering, Chung-Ang University, Seoul 06973, Korea
* Correspondence: hongsyun@kaist.ac.kr (H.Y.); paiktae@cau.ac.kr (T.P.);
Tel.: +82-042-350-3975 (H.Y.); +82-02-820-5435 (T.P.)

Received: 31 July 2019; Accepted: 26 August 2019; Published: 1 September 2019

Abstract: The self-assembly of colloidal inorganic nanocrystals (NCs) offers tremendous potential for the design of solution-processed multi-functional inorganic thin-films or nanostructures. To date, the self-assembly of various inorganic NCs, such as plasmonic metal, metal oxide, quantum dots, magnetics, and dielectrics, are reported to form single, binary, and even ternary superlattices with long-range orientational and positional order over a large area. In addition, the controlled coupling between NC building blocks in the highly ordered superlattices gives rise to novel collective properties, providing unique optical, magnetic, electronic, and catalytic properties. In this review, we introduce the self-assembly of inorganic NCs and the experimental process to form single and multicomponent superlattices, and we also describe the fabrication of multiscale NC superlattices with anisotropic NC building blocks, thin-film patterning, and the supracrystal formation of superlattice structures.

Keywords: BNSL; superlattice; self-assembly; colloidal nanocrystal; binary nanocrystal superlattice

---

1. Introduction

In the past decades, colloidal inorganic nanocrystals (NCs) have received considerable attention in several research fields because of their interesting size-dependent properties, such as their quantum confinement effect and localized surface plasmonic effect, which are not be observed in the bulk [1,2]. With extensive research efforts, there has been significant progress in the development of synthetic methods for inorganic NCs, enabling the precise tuning of their size, chemical composition, crystallinity, and shape, which is very important in controlling their properties. In addition to rendering unique material properties to individual NCs, the self-assembly of NCs provides a "bottom-up" approach for the fabrication of micro- or macroscale NC-based films with highly complicated nanostructures, which are difficult to achieve through conventional lithography-based fabrication processes. Moreover, NC building blocks enable solution-based, cheap, and scalable processes, which are highly beneficial for industrial applications.

One of the most interesting applications of NC building blocks is the colloidal self-assembly of NCs into ordered crystalline structures; that is, NC building blocks form various crystal structures, including face-centered cubic (fcc), body-centered cubic (bcc), and hexagonal close packed (hcp) structures, similar to how atoms or ions form crystalline structures [3,4]. More importantly, the use of two different types of NCs can yield highly ordered binary NC superlattices (BNSLs) with various packing structures, such as NaCl, $MgZn_2$, and $NaZn_{13}$, depending on the size and concentration ratio [5]. BNSLs exhibit not only structural diversity for tuning NC–NC interactions through the choice of NC constituents or packing symmetry, but also collective properties that are distinct from the sum of the individual characteristics. It is highly important to understand the self-assembly behavior of NCs for the development of novel materials because of their superior controllability in material design. In this review, we will broadly describe the self-assembly of colloidal NCs, including the fabrication method, formation mechanism of self-assembly, structural diversity of BNSLs, and mesoscale structure of self-assembled NC superlattices.

2. Self-Assembly of NCs

2.1. Methods for the Self-Assembly of NCs

Highly ordered NC superlattices can be prepared by the self-assembly of colloidal NCs. Mostly, colloidal inorganic NCs are synthesized by the high-temperature solvothermal decomposition process, which yields highly monodispersed NCs coated with alkyl chain ligands [6–8]. A few approaches have been adopted to build highly ordered NC superlattices with the prepared NC building blocks. One of the methods is simply drop-casting colloidal NCs in non-polar solvents such as hexane, toluene, or chloroform onto a solid substrate and allowing them to dry for a couple of minutes [9]. When the solvent evaporates, the NCs are densely solidified through various kinds of interactions including NC–NC interactions (i.e., van der Waals force and electrostatic interaction) and ligand–ligand interactions (i.e., hydrogen bonding). In addition to the simple drop-casting method, the recrystallization method has been used for the preparation of three-dimensional (3D) ordered NC superlattices by using a polar solvent to destabilize NC dispersion in a non-polar solvent, as shown in Figure 1a–c [10,11]. As the non-polar solvent slowly evaporates from the QD dispersion in the presence of the polar solvent, QDs start crystallizing because of their decreased solubility in the solution. Consequently, 3D micro-sized NC superlattices are formed because of van der Waals interactions between the NCs and the change in free energy during the crystallization process.

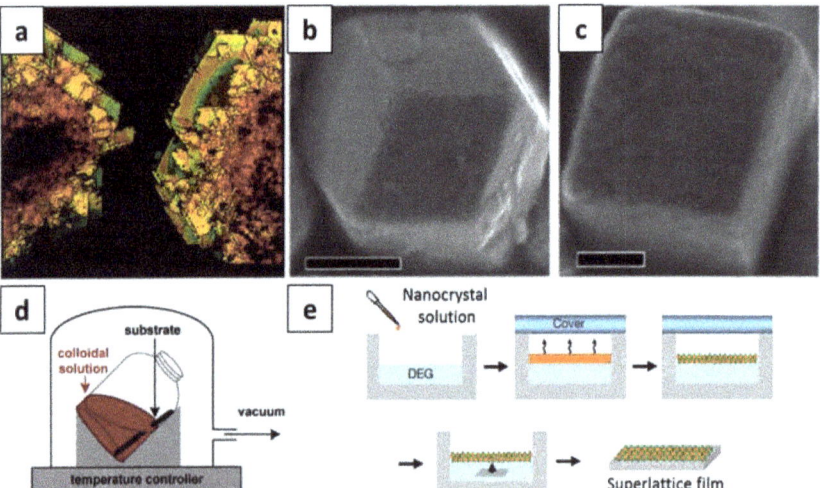

Figure 1. (a) Dark field optical micrograph of colloidal crystals formed by 2 nm CdSe nanocrystals (NCs) (Reproduced with permission from [10], Copyright American Association for the Advancement of Science, 1995). Scanning electron microscope (SEM) images of self-assembled supercrystals of (b) octahedral and (c) cubic Pt NCs (Reproduced with permission from [11]. Copyright American Chemical Society, 2013). Schematic illustration of the self-assembly of NCs by (d) the slow evaporation of NC solution under vacuum (Reproduced with permission from [12], Copyright American Chemical Society, 2006). and (e) liquid–air interface assembly (Reproduced with permission from [13]. Copyright Springer Nature, 2010). Scale bars in Figure 1b,c represent 500 nm and 200 nm, respectively.

Another method for the self-assembly of NCs involves the slow evaporation of the solvent from the NC solution, inducing the crystallization of the NCs [12]. As shown in Figure 1d, a substrate (e.g., transmission electron microscope (TEM) grid or silicon wafer) is placed in a container with the NC solution. Then, the container is placed in a chamber and tilted by 60–70°. Next, the solvent is slowly evaporated under a low-pressure vacuum at 45 °C. As the concentration of NCs in the solution increases, it reaches the solubility limit of NCs in the solution, leading to the crystallization of NCs. Consequently, a well-ordered NC superlattice is formed on the substrate. A similar approach has been adopted to obtain NC superlattices at the liquid–air interface [13,14]. When NCs in non-polar solvents such as hexane and toluene are drop-casted on top of an immiscible polar solvent (e.g., ethylene glycol and diethylene glycol) in a well, followed by covering the top with a slide glass, the non-polar solvent on top of the liquid substrate slowly evaporates, and as the concentration increases, NCs are crystallized. Finally, a thin, long-range ordered NC film forms on top of the polar solvent, which is then transferred to a solid substrate for characterization. An advantage of the liquid–air interface self-assembly technique is that it yields uniform NC superlattice thin-films over a large area within a short time.

2.2. Self-Assembly of Spherical NCs

When self-assembled, NCs form highly ordered superlattices, resembling the atomic crystal structure; that is, self-assembled NCs can exhibit various packing symmetries including fcc and hcp, which have the highest packing density (74%), and non-close packing symmetries such as bcc and simple cubic (sc) symmetries with packing densities of 68% and 52%, respectively. When hard spheres assemble, preferably, packing occurs with the highest free volume entropy, leading to close-packed symmetries (i.e., fcc and hcp). Nonetheless, non-close packing symmetries, such as bcc and sc, are also often observed for NC superlattices, which is hard to explain on the basis of the entropy-driven assembly mechanism. Experimentally, it has been found that the softness of NCs, λ, which is defined as the extended ligand length-to-core radius ratio, is a very important factor that must be considered in the determination of the packing symmetries of NC superlattices [15]. This can be attributed to the interplay between entropic and enthalpic effects in fcc and bcc symmetries [16–18]. When λ is small—or in other words, when the ligand length is relatively short compared with the core radius—NCs act as hard spheres and preferably adopt an fcc symmetry due to its larger enthalpic gain compared to that of bcc. On the other hand, when λ is relatively large, NCs adopt a bcc symmetry because the entropic effects from ligand packing become dominant.

To interpret the softness-dependent self-assembly behavior of NCs, various kinds of theoretical models have been proposed. For example, on the basis of space filling between inorganic NC cores, two different models have been proposed: the optimal packing model (OPM) [19] and overlap cone model (OCM) [20]. The OPM postulates the densest packing of organic ligands along the NC core-to-core axis. The self-assembly behavior of NCs can be successfully predicted with the OPM using λ as a variable. For example, according to the OPM, the effective radius (r_i) of NCs, which is half of the core-to-core distance, can be expressed as $r_i = R(1 + 3\varepsilon\lambda)^{\frac{1}{3}}$, where R is the inorganic NC core radius, ε is the ratio of the maximum surface area occupied by a ligand to the actual surface area covered, and λ is the ratio between the extended chain length to the core radius [19,21]. This formula can successfully describe not only the NC–NC separation distance but also the λ required for the transition from fcc to bcc. While the OPM fits three-dimensionally assembled NC superlattices well, the NC–NC separation determined using the OPM formula cannot be well applied to low-coordinated NC superlattices. To correct this flaw, the OCM has been proposed, wherein the truncated ligand cones intersect with each other and maximize the packing density, resulting in a shorter interparticle distance than that in the OPM. As shown in Figure 2, the OCM can successfully predict the interparticle distance between NCs in low coordination.

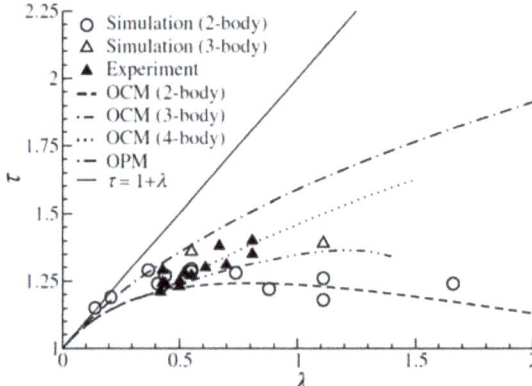

Figure 2. Scaled equilibrium distance τ vs. λ. When the surface ligands do not overlap each other, $\tau = 1 + \lambda$. Therefore, as the points or lines are more away from the solid line ($\tau = 1 + \lambda$), it indicates more overlapped surface ligands. (Reproduced with permission from [20]. Copyright American Institute of Physics, 2009). OCM: Overlap cone model. OPM: Optimal packing model.

Later, the difference between the OPM and OCM was explained by Boles and Talapin, and they attributed it to the many-body effect in NC superlattices [22]; that is, because soft organic ligands can be deformed when there is another NC nearby, the NC–NC distance becomes smaller when NCs are surrounded by a low number of NCs. On the other hand, NCs in high coordination exhibit a longer NC–NC separation distance because of the limited deformability of the ligands, which shows good agreement with the OPM rather than with the OCM. The presence of the many-body effect in NC superlattices has led to the development of the orbifold topological model (OTM), which treats the deformable ligand coronas as topological defects [23]. The OTM predicts the formation behavior of BNSLs, the NC separation distance, and the stability of packing symmetries well.

Alkyl-chain based ligands on NC surfaces have been replaced by other organic materials to diversify the phase diagram of NC superlattices. For example, polymeric ligands have been reported to offer enriched NC packing symmetries [24] as well as enhanced mechanical stability [25]. Moreover, polymeric ligand-coated NCs show a different self-assembly manner from that of conventional alkyl chain-coated NCs. For example, it is well-known that alkyl-chain-coated NCs preferentially adopt a bcc symmetry when λ is over 0.6–0.7 [15,16,19]. On the other hand, Yun et al. recently reported that Au@PS nanoparticles adopt fcc packing symmetries even at a λ of 3.0, which was attributed to the grafting density effect, wherein ligand penetration is limited around the NC surface, lowering the "effective softness" of the nanoparticles and thereby leading to the formation of assemblies with fcc symmetries. [26] The authors formulated the "effective softness", λ_{eff}, as a function of grafting density by including the concentrated polymer brush (CPB) regime as a part of the "hard core", which was then applied to the OPM and successfully predicted the effective nanoparticle (NP) radius more accurately than the prediction by λ, as shown in Figure 3. Self-assemblies of DNA-coated NCs have been widely demonstrated [27,28]. DNA-based ligands can be designed to control the ligand–ligand interaction, which enables the programmable self-assembly of NCs. Dendrimers can also provide a wide range of interparticle spacing by changing the dendritic generation grafted on the NC surface [29,30].

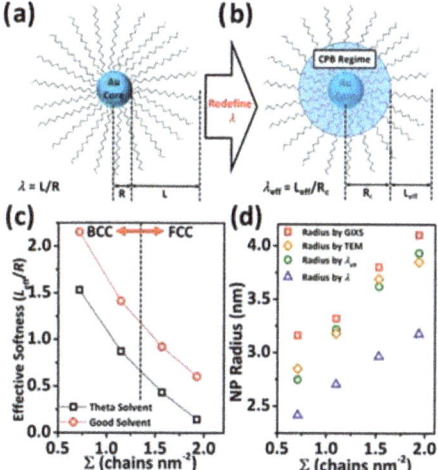

Figure 3. Illustrations of (**a**) the conventional concept of softness (λ) of nanoparticles (NPs) and (**b**) the concept of effective softness (λ_{eff}). (**c**) Effective softness variation as a function of grafting density (Σ). (**d**) Comparison between the effective NP radius predicted by λ (blue triangle), λ_{eff} (green circle), and experimental results obtained from transmission electron microscopy (TEM, orange diamond) and grazing incidence x-ray scattering (GIXS, red square) (Reproduced with permission from [26]. Copyright American Chemical Society, 2019).

2.3. Self-Assembly of BNSLs

In addition to single-component NC superlattices, highly ordered NC superlattices, called BNSLs, can be formed by using either a single type of NCs or a mixture of two different types of NCs with different sizes [31], shapes [5], and properties [32]. For example, Redl et al. demonstrated the formation of BNSLs using γ-Fe$_2$O$_3$ (magnetic) and PbSe (semiconducting) NCs with a precisely controlled size and narrow size distribution [33]. By varying the size ratio and concentration ratio of the NCs, they prepared BNSLs with various packing symmetries including AB$_2$, AB$_5$, and AB$_{13}$. When the size ratio ($d_{PbSe}/d_{\gamma\text{-Fe2O3}}$) was 0.58, AB$_2$ and AB$_{13}$ BNSLs were formed, and at a higher value of 0.63, AB$_5$ BNSL was formed. This was a particularly interesting BNSL system because two different NCs with independently tunable optical (PbSe) and magnetic (γ-Fe$_2$O$_3$) characteristics were employed, which could enable the fine tuning of material properties. More importantly, BNSLs have enormous structural diversity, as demonstrated by Shevchenko et al., wherein 15 different BNSL symmetries, such as NaCl-, CuAu-, MgZn$_2$-, MgNi$_2$-, AlB$_2$-, Cu$_3$Au-, CaCu$_5$-, and NaZn$_{13}$-type (Figure 4), were observed [5]. BNSLs with various symmetries could be prepared using several types of NCs including Au, PbSe, Pd, Ag, and γ-Fe$_2$O$_3$ with different sizes.

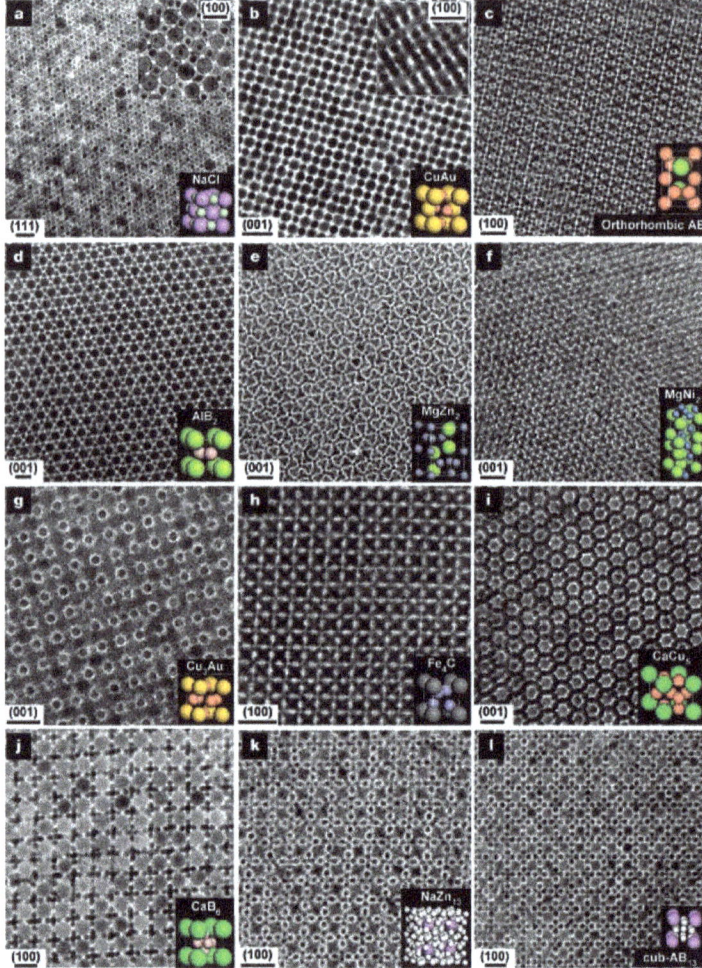

Figure 4. Structural diversity of binary nanocrystal superlattices (BNSLs). (**a**) 13.4 nm g-Fe2O3 and 5.0 nm Au NCs (NaCl-type), (**b**) 7.6 nm PbSe and 5.0 nm Au NCs (CuAu-type), (**c**) 6.2 nm PbSe and 3.0 nm Pd (AB-type), (**d**) 6.7 nm PbS and 3.0 nm Pd (AlB$_2$-type), (**e**) 6.2 nm PbSe and 3.0 nm Pd (MgZn$_2$-type), (**f**) 5.8 nm PbSe and 3.0 nm Pd (MgNi$_2$-type), (**g**) 7.2 nm PbSe and 4.2 nm Ag (Cu$_3$Au-type), h) 6.2 nm PbSe and 3.0 nm Pd (Fe$_4$C-type), i) 7.2 nm PbSe and 5.0 nm Au (CaCu$_5$-type), (j) 5.8 nm PbSe and 3.0 nm Pd (CaB$_6$-type), (**k**) 7.2 nm PbSe and 4.2 nm Ag (NaZn$_{13}$-type), and (**l**) 6.2 nm PbSe and 3.0 nm Pd (cub-AB$_{13}$-type). Scale bars represent 20 nm (**a–c,e,f,i–l**) and 10 nm (**d,g,h**). (Reproduced with permission from [5]. Copyright Springer Nature, 2006).

The structure of BNSL thin-films are relatively stable in ambient conditions. There are studies which report that the thermal stability of NCs can be significantly enhanced upon the formation of BNSL compared with that of single-component NC films. For example, although FePt NCs are thermally unstable and easily sintered [34], it was reported that the BNSL structure consisting of FePt and MnO was preserved even after thermal annealing at 650 °C [35]. This could be attributed to the fact that the presence of thermally stable MnO NCs around FePt NCs spatially confine them to prevent coalescence. In addition to the thermal stability, the mechanical stability of BNSLs has been demonstrated, showing that the BNSLs can form free-standing membranes [13] and even monolayers [36]. Also, the membranes are robust enough to enable pattern transfer [37].

To understand the formation mechanism of BNSLs, the change in the free energy of the system must be taken into account. The free energy change is determined by NC–NC interactions as well as the entropic change during the formation of BNSLs. In addition, as reported, during the formation of opal crystals by micron-sized colloidal particles, the NCs self-assemble into well-ordered superlattices even without the presence of NC–NC interactions, which is called "entropy-driven self-assembly". When colloidal particles form well-ordered arrays with the highest packing fraction, the system obtains an additional free volume, which eventually leads to maximum entropy. In addition to the entropy-based principle, the "space-filling principle", which was proposed by Murray and Sanders [38], can be used to describe the self-assembly behavior of BNSLs. For single-component hard spheres, the highest packing symmetry is either fcc or hcp (both have a filling fraction of 0.74). When NCs with two different sizes are assembled, BNSLs are formed with a packing fraction of over 0.74, which is thermodynamically more stable than close-packed symmetries (i.e., fcc and hcp). Therefore, according to the space-filling principle, targeted BNSL structures can be obtained by tuning the size ratio between two different NCs. Figure 5 presents the phase diagram of BNSLs based on the space-filling principle. As observed, each BNSL symmetry shows different packing fractions. Moreover, it has been experimentally demonstrated that two different NCs can assemble into BNSLs with packing fractions higher than those of close-packed structures (0.74), as shown in the phase diagram.

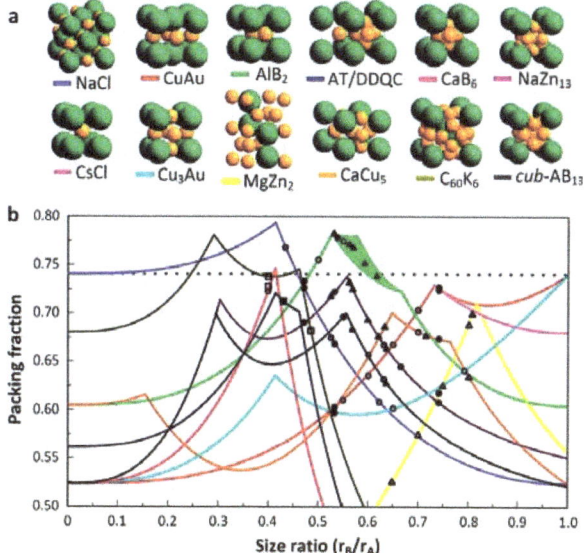

Figure 5. (a) Illustrations of different types of BNSL packing symmetries. (b) Phase diagram showing the packing fraction of BNSLs as a function of size ratio. (Reproduced with permission from [22]. Copyright American Chemical Society, 2015).

It has been reported that BNSLs with both translational and rotational orders, and with rotational order but without translational order, can be formed, which are called quasicrystals [39]. For example, when a combination of 5 nm Au NCs and 13.4 nm Fe_2O_3 NCs was used for self-assembly, the NCs formed BNSLs with a 12-fold rotational symmetry but without any translational symmetry (dodecagonal quasicrystals). In this case, the size ratio was 0.34, wherein the CaB_6 and AlB_2 BNSLs have the same packing fraction, indicating that both structures are thermodynamically stable. AlB_2 and CaB_6 BNSLs consist of triangular and square tiles, respectively, whereas quasicrystal structures contain periodic arrays of triangular and square tiles. Therefore, when the size ratio of 0.34 was used, dodecagonal quasicrystals were formed through the periodic arrangement of BNSLs with both symmetries (AlB_2 and CaB_6).

In addition to BNSLs, more complicated NC superlattices can be formed using more than two types of NCs. For instance, in 2009, Vanmaekelbergh et al. demonstrated the formation of well-ordered ternary NC superlattices using 12.1 nm PbSe NCs (A), 7.9 nm PbSe NCs (B), and 5.8 CdSe NCs (C): ABC_4 (isostructural with $AlMgB_4$) along with AB_2 (AlB_2) and BC_2 ($MgZn_2$) [40]. The detailed structure of the ternary NC superlattice was confirmed by 3D electron tomography, as shown in Figure 6. Interestingly, the packing fraction of the $AlMgB_4$ ternary NC superlattice was 0.64, which was lower than those of AlB_2 (0.76) and $MgZn_2$ (0.67). This was attributed to the fact that the ternary NC superlattice was obtained not by the effects of entropy but rather by the combination of NC–NC interactions and thermodynamic factors. To predict the most stable packing symmetry of NC superlattices, the total energy must be calculated.

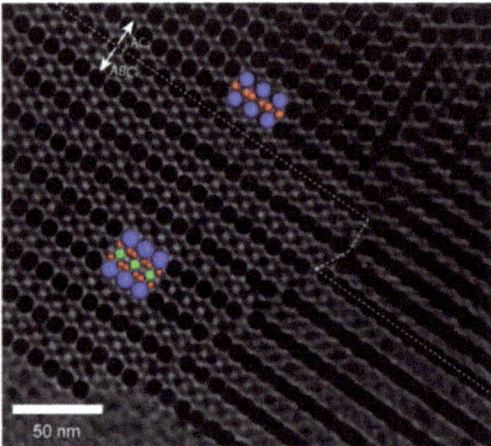

Figure 6. Transmission electron microscopy (TEM) image of the $AlMgB_4$-type ternary superlattice (PbSe(l)-PbSe(m)-CdSe(s)$_4$ nanocrystal superlattice, l = large, m = medium, and s = small) in epitaxial contact with the AlB_2-type binary superlattice (PbSe(l)-CdSe(s)$_2$). The TEM image and schematic show the (100) planes of the ternary superlattice, in which PbSe (l, blue spheres), PbSe (m, green spheres), and CdSe (s, red spheres) can be individually observed. (Reproduced with permission from [40]. Copyright John Wiley and Sons, 2009.

Generally, BNSLs are grown as 2D thin films, depending on the fabrication method. In 2015, Murray and Kagan et al. reported the fabrication of multiscale-patterned BNSLs by colloidal self-assembly and transfer printing [37]. BNSL thin films are first fabricated from NC building blocks of various materials such as metals, semiconductors, magnetics, and dielectrics by the liquid–air interfacial assembly method; the BNSL thin films formed at the interface are then transferred onto patterned polydimethylsiloxane (PDMS) molds by the Langmuir–Schaefer technique. During this process, only the BNSL thin film on the raised region of the PDMS pattern is transferred; thus, patterned BNSL nanostructures are obtained on the substrate. The transferred structures exhibit a mesoscale order, while the BNSLs maintain the nanoscale order, resulting in multiscale hierarchical architectures. Figure 7 shows the TEM and SEM images of a patterned BNSL film obtained by self-assembly and transfer printing. The SEM images reveal the formation of mesoscale line patterns. The TEM images reveal that the AlB_2-type BNSL of Au and FeOx NCs is maintained after transfer printing, exhibiting long-range order over a large area. The nanoscale BNSL structures can be readily tuned by changing the size and composition of the NC building blocks, as previously described. In addition, the mesoscale pattern (circular or square arrays) of BNSLs can be tailored by changing the shape of the PDMS molds. Moreover, the BNSL patterns can be stacked by layer-by-layer transfer printing, indicating that the fabrication of complex BNSL structures is possible by sequential self-assembly and multiple transfer printing.

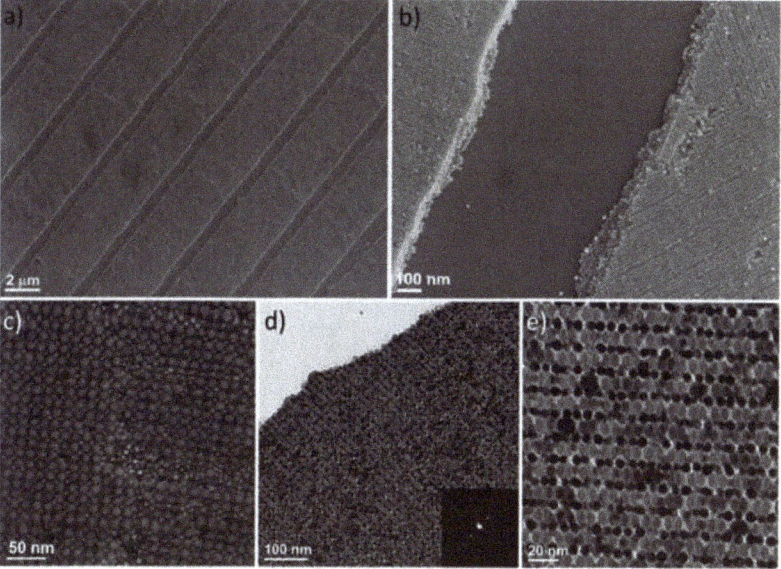

Figure 7. Hierarchical BNSL nanostructures formed by liquid–air interfacial assembly and transfer printing. (**a**–**c**) SEM images, and (**d**) low-magnification (inset: selected-area electron diffraction pattern) and (**e**) high-magnification TEM images of patterned AlB$_2$ BNSLs assembled from FeOx and Au NCs (Reproduced with permission from [37]. Copyright American Chemical Society, 2017).

BNSL can be formed not only into 2D thin films but also into a 3D confined emulsion. For example, Wang et al. reported the formation of colloidal BNSL supracrystals with various symmetries, as shown in Figure 8 [41]. They demonstrated the preparation of BNSL supracrystals by oil-in-water emulsion droplets of Au and Fe$_3$O$_4$ NCs followed by the evaporation of the oil phase, which induces the co-crystallization of NCs into 3D confined BNSL structure. By controlling the ratio between Au and Fe$_3$O$_4$ NCs, various BNSL symmetries such as AB$_2$, AB$_3$, and AB$_{13}$-type symmetries could be achieved. Recently, such 3D confined BNSL supracrystals of CoFe$_2$O$_4$-Fe$_3$O$_4$ have been reported to have superior lithium storage properties compared with their single-component counterparts [42]. The enhanced electrochemical properties were attributed to the non-close packed symmetry of the BNSL supracrystals, which promoted better mass transport as well as endurance against volumetric changes during lithiation and delithiation processes.

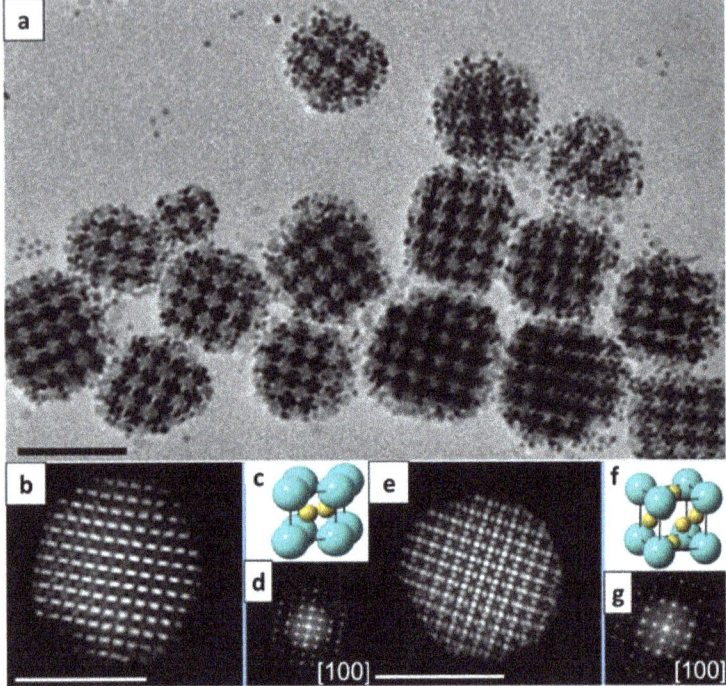

Figure 8. (a) A TEM image of ico-NaZn$_{13}$-type supracrystals. High-angle annular dark field scanning TEM (HAADF-STEM) images of (b) AlB$_2$-type supracrystals and (e) AuCu$_3$-type supracrystals. (c,f) The unit cells of the AlB$_2$-type lattice and AuCu$_3$-type lattice, and (d,g) their fat Fourier transformation (FET) patterns. The scale bar in all the TEM images is 100 nm. (Reproduced with permission from [41]. Copyright American Chemical Society, 2018)

2.4. Self-Assembly of Anisotropic BNSLs

Various types of anisotropic NCs including rods, cubes, tetrahedrons, and plates have been reported. Anisotropic NCs have received particular attention in optics [43], magnetics [44], catalysis [45], and electronics. [46] Accordingly, considerable efforts have been made to prepare NC superlattices using anisotropic NCs as building blocks, thereby maximizing the potential application of NCs. CdSe nanorods are one of the most extensively studied anisotropic NCs for self-assembly because of their shape-dependent light polarization properties. According to Talapin et al., when 1D CdSe nanorods are self-assembled by controlling the dispersibility of nanorods, either nematic or smectic liquid crystals are formed. [47] The formation of a long-range nanorod superstructure was attributed to a combination of strong side-to-side van der Waals interactions, antiparallel side-by-side dipole pairing, and entropy effects, which led to an increase in the free volume space, thereby achieving the highest packing density. The self-assembly of anisotropic NCs of various materials including Au, Cu$_x$S, LaF$_3$, β-NaYF$_4$, and GdF$_3$ has been reported [48–52].

In addition to the superlattices of single components, the formation of BNSLs of multicomponent anisotropic NCs has also been reported. For instance, in 2006, the formation of BNSLs was demonstrated using LaF$_3$ triangular nanoplates (9 nm side) and 5 nm spherical Au NCs [5]. Also, AB$_2$-type BNSLs could be also fabricated by the combination of Fe$_3$O$_4$ NCs and spherical β-NaYF$_4$ nanorods as building blocks [53]. The experimentally obtained results were compared with Monte Carlo simulation results, and it was found that the ligand–ligand interaction and depletion attraction caused by the extra ligands around the NCs affected the formation of BNSLs of anisotropic NCs. In addition, on the basis of the space-filling principle and size ratio-dependent self-assembly behavior, it was found that entropy-driven free energy maximization determined the BNSL symmetry, leading to the formation of BNSL structures with the highest packing density.

Moreover, BNSLs consisting of two different anisotropic NCs, LaF$_3$ nanodisks (2 nm thickness and 15–25 nm diameter) and CdSe-CdS nanorods, have been reported [54]. In this work, NCs dispersed in hexane were drop-casted on top of diethylene glycol (immiscible with hexane) and slowly dried at the liquid–air interface, yielding highly ordered BNSL structures. In particular, because of the shape anisotropy of these NCs, they showed shape-selective interactions. The nanodisks self-assembled to form a stacked columnar structure through face-to-face van der Waals interactions, yielding a 2D hexagonally packed liquid crystalline structure, as presented in Figure 9a. On the other hand, the 1D nanorods assembled to form smectic lamellar liquid crystalline structures via side-by-side van der Waals interactions (Figure 9b). Moreover, during the self-assembly of the two different anisotropic NCs, the nanorods vertically aligned to fill the interstitial sites between the hexagonally packed columnar structure of the nanodisks, filling the hierarchically assembled BNSL structure, as shown in Figure 9c. It was found that AB-, AB$_2$-, and AB$_6$-type BNSL structures could be formed depending on the size and concentration ratio between the nanodisks and nanorods. This phenomenon further confirms that the BNSL structure is formed to maximize the packing fraction of NCs in the system, similar to the formation principle of spherical BNSL structures. This reveals a particularly interesting aspect of the formation of anisotropic NC-based BNSLs: the entropic factor affects the final packing symmetry of BNSLs of anisotropic NCs even while the orientation is controlled. This result indicates that the formation of BNSLs of anisotropic NCs is more complicated; therefore, the directional order, van der Waals interactions between the NCs, and entropic effects must be taken into consideration. In addition to BNSLs, ternary NC superlattices comprising three different anisotropic NCs (i.e., nanorods and two different nanodisks) were reported by the authors.

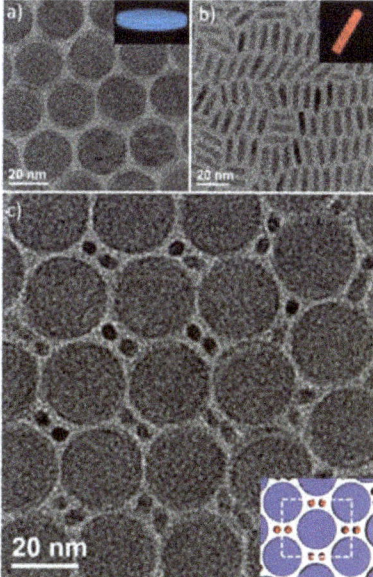

Figure 9. TEM images of binary superlattices of two anisotropic NC building blocks. (**a**) LaF$_3$ nanodisks, (**b**) CdSe/CdS nanorods, and (**c**) AB$_2$ BNSLs of LaF$_3$ nanodisks and CdSe/CdS nanorods. Inset is the illustration of the BNSL structure. Reproduced with permission from [54]. Copyright American Chemical Society, 2015.

Interestingly, anisotropic NCs can be self-assembled through selective interaction, depending on their shape, in a similar way to puzzles, which was previously demonstrated in lock and key colloids. In 2013, a shape-complementary BNSL structure comprising GdF_3 rhombic nanoplates and Gd_2O_3 tripodal nanoplates was reported [55]. In this case, the side length and interior angle of the rhombic nanoplates were precisely controlled to be close to those of the tripodal nanoplates. Subsequently, the two different anisotropic NCs were assembled into interlocked BNSLs on the basis of shape complementarity. This work demonstrates that the self-assembly of shape-complementary anisotropic building blocks may provide a unique design rule to direct the formation of BNSL thin films over a large area with high complexity in a predictable way.

3. Perspectives

The self-assembly of colloidal NCs offers significant potential for sophisticated, novel material design because of their structural diversity and the variety of material choices, as well as the formation of complicated structures. Since NC–NC interactions can be more effectively tuned through the formation of various types of BNSL structures than those of single-component NC superlattices, a larger variety of novel material characteristics can be achieved. Many reports have been published describing the methods of formation, the mechanism, and the structural characterization of NC superlattices. Although structure dependent, synergistic collective interactions between NCs in BNSLs are reported in electronics [56–58], optics [59], catalysis [32], and magnetics [56]; however, a limited number of studies on the intrinsic material properties of highly ordered NC superlattices, particularly the BNSLs, have been performed so far.

To achieve the full potential of BNSL-based materials, it is necessary to obtain an in-depth understanding of complicated NC–NC interactions within various BNSL structures. For example, some NC–NC interactions can occur over a few nanometers, while energy or electron transfer between NCs exponentially decreases due to the presence of surface ligands on NC surfaces. Therefore, it is important to reduce the distance between NCs while preserving the BNSL structures in order to enhance the interaction between NCs. The development of a fabrication method of BNSLs using NCs which their ligands are stripped or exchanged to short-chain ligands will enable BNSLs to be applied in many emerging fields. Moreover, through the multiscale fabrication method of BNSLs, synergistic effects between mesoscale structural effects and collective properties of BNSLs may be realized. For examples, mesoscale patterns of BNSLs may add mesoscale photonic effects to the collective properties of BNSLs. In addition, BNSL supracrystals may be utilized in the application of biomedical imaging agents.

In terms of processing, it is imperative to develop large-area-BNSL formation techniques for the commercialization of BNSL-based novel materials. Although the facile and high-throughput fabrication method of large-area BNSL structures are demonstrated [60], there are many experimental difficulties to forming large-area, uniform superlattice thin-films. Therefore, the undertaking of systematic studies to understand the effect of self-assembly conditions, including the choices of the building blocks and the concentration of NCs on the formation of defect-free, long-range ordered BNSLs, would be important. If the fabrication of uniform large-area NC thin films can be realized, more accurate material analysis can be performed to identify BNSLs with outstanding material properties, eventually promoting the application of NC thin films for the development of novel materials in several industrial fields.

Author Contributions: Investigation, H.Y., and T.P.; writing—review and editing, H.Y., and T.P.; supervision, H.Y., and T.P.; funding acquisition, T.P.

Funding: This research was supported by Creative Materials Discovery Program through the National Research Foundation of Korea (NRF) funded by Ministry of Science and ICT (NRF-2018M3D1A1059001). This research was supported by the Chung-Ang University Research Grants, 2018.

Conflicts of Interest: The authors declare no conflict of interest.

References

1. Kagan, C.R. Flexible colloidal nanocrystal electronics. *Chem. Soc. Rev.* **2019**, *48*, 1626–1641. [CrossRef] [PubMed]
2. Hsu, S.-W.; Rodarte, A.L.; Som, M.; Arya, G.; Tao, A.R. Colloidal Plasmonic Nanocomposites: From Fabrication to Optical Function. *Chem. Rev.* **2018**, *118*, 3100–3120. [CrossRef] [PubMed]
3. Murray, C.B.; Kagan, C.R.; Bawendi, M.G. Synthesis and characterization of monodispersed nanocrystals and close-packed nanocrystals assemblies. *Annu. Rev. Mater. Sci.* **2000**, *30*, 545–610. [CrossRef]
4. Pileni, M.P. Supracrystals of inorganic nanocrystals: An open challenge for new physical properties. *Acc. Chem. Res.* **2008**, *41*, 1799–1809. [CrossRef] [PubMed]
5. Shevchenko, E.V.; Talapin, D.V.; Kotov, N.A.; O'Brien, S.; Murray, C.B. Structural diversity in binary nanoparticle superlattices. *Nature* **2006**, *439*, 55–59. [CrossRef] [PubMed]
6. Murray, C.B.; Norris, D.J.; Bawendi, M.G. Synthesis and characterization of nearly monodisperse CdE (E= S, Se, Te) semiconductor nanocrystallites. *J. Am. Chem. Soc.* **1993**, *115*, 8706–8715. [CrossRef]
7. Peng, X.; Manna, L.; Yang, W.; Wickham, J. Shape control of CdSe nanocrystals. *Nature* **2000**, *404*, 59–61. [CrossRef] [PubMed]
8. Park, J.; An, K.; Hwang, Y.; Park, J.-G.; Noh, H.-J.; Kim, J.-Y.; Park, J.-H.; Hwang, N.-M.; Hyeon, T. Ultra-large-scale syntheses of monodisperse nanocrystals. *Nat. Mater.* **2004**, *3*, 891–895. [CrossRef]
9. Connolly, S.; Fullam, S.; Korgel, B.; Fitzmaurice, D. Time-resolved small-angle X-ray scattering studies of nanocrystal superlattice self-assembly. *J. Am. Chem. Soc.* **1998**, *120*, 2969–2970. [CrossRef]
10. Murray, C.; Kagan, C.; Bawendi, M. Self-organization of CdSE nanocrystallites into three-dimensional quantum dot superlattices. *Science* **1995**, *270*, 1335–1338. [CrossRef]
11. Kang, Y.; Li, M.; Cai, Y.; Cargnello, M.; Diaz, R.E.; Gordon, T.R.; Wieder, N.L.; Adzic, R.R.; Gorte, R.J.; Stach, E.A.; et al. Heterogeneous catalysts need not be so "Heterogeneous": Monodisperse Pt nanocrystals by combining shape-controlled synthesis and purification by colloidal recrystallization. *J. Am. Chem. Soc.* **2013**, *135*, 2741–2747. [CrossRef] [PubMed]
12. Shevchenko, E.V.; Talapin, D.V.; Murray, C.B.; O'Brien, S. Structural characterization of self-assembled multifunctional binary nanoparticle superlattices. *J. Am. Chem. Soc.* **2006**, *128*, 3620–3637. [CrossRef] [PubMed]
13. Dong, A.; Chen, J.; Vora, P.M.; Kikkawa, J.M.; Murray, C.B. Binary nanocrystal superlattice membranes self-assembled at the liquid-air interface. *Nature* **2010**, *466*, 474–477. [CrossRef] [PubMed]
14. Bigioni, T.P.; Lin, X.M.; Nguyen, T.T.; Corwin, E.I.; Witten, T.A.; Jaeger, H.M. Kinetically driven self assembly of highly ordered nanoparticle monolayers. *Nat. Mater.* **2006**, *5*, 265–270. [CrossRef] [PubMed]
15. Korgel, B.; Fitzmaurice, D. Small-angle x-ray-scattering study of silver-nanocrystal disorder-order phase transitions. *Phys. Rev. B* **1999**, *59*, 14191–14201. [CrossRef]
16. Goodfellow, B.W.; Yu, Y.; Bosoy, C.A.; Smilgies, D.M.; Korgel, B.A. The role of ligand packing frustration in body-centered cubic (bcc) superlattices of colloidal nanocrystals. *J. Phys. Chem. Lett.* **2015**, *6*, 2406–2412. [CrossRef]
17. Kaushik, A.P.; Clancy, P. Solvent-driven symmetry of self-assembled nanocrystal superlattices—A computational study. *J. Comput. Chem.* **2013**, *34*, 523–532. [CrossRef]
18. Zha, X.; Travesset, A. Stability and free energy of nanocrystal chains and superlattices. *J. Phys. Chem. C* **2018**, *122*, 23153–23164. [CrossRef]
19. Landman, U.; Luedtke, W.D. Small is different: Energetic, structural, thermal, and mechanical properties of passivated nanocluster assemblies. *Faraday Discuss.* **2004**, *125*, 1–22. [CrossRef]
20. Schapotschnikow, P.; Vlugt, T.J.H. Understanding interactions between capped nanocrystals: Three-body and chain packing effects. *J. Chem. Phys.* **2009**, *131*, 124705. [CrossRef]
21. Travesset, A. Soft skyrmions, spontaneous valence and selection rules in nanoparticle superlattices. *ACS Nano* **2017**, *11*, 5375–5382. [CrossRef] [PubMed]
22. Boles, M.A.; Talapin, D.V. Many-body effects in nanocrystal superlattices: Departure from sphere oacking explains stability of binary phases. *J. Am. Chem. Soc.* **2015**, *137*, 4494–4502. [CrossRef]
23. Henzie, J.; Andrews, S.C.; Ling, X.Y.; Li, Z.; Yang, P. Oriented assembly of polyhedral plasmonic nanoparticle clusters. *Proc. Natl. Acad. Sci. USA* **2013**, *110*, 6640–6645. [CrossRef] [PubMed]

24. Ye, X.; Zhu, C.; Ercius, P.; Raja, S.N.; He, B.; Jones, M.R.; Hauwiller, M.R.; Liu, Y.; Xu, T.; Alivisatos, A.P. Structural diversity in binary superlattices self-assembled from polymer-grafted nanocrystals. *Nat. Commun.* **2015**, *6*, 10052. [CrossRef] [PubMed]
25. Gu, X.W.; Ye, X.; Koshy, D.M.; Vachhani, S.; Hosemann, P.; Alivisatos, A.P. Tolerance to structural disorder and tunable mechanical behavior in self-assembled superlattices of polymer-grafted nanocrystals. *Proc. Natl. Acad. Sci. USA* **2017**, *114*, 2836–2841. [CrossRef]
26. Yun, H.; Yu, J.; Lee, Y.J.; Kim, J.-S.; Park, C.H.; Nam, C.; Han, J.; Heo, T.-Y.; Choi, S.-H.; Lee, D.C.; et al. Symmetry transitions of polymer-grafted nanoparticles: Grafting density effect. *Chem. Mater.* **2019**, *31*, 5264–5273. [CrossRef]
27. Kim, Y.; Macfarlane, R.J.; Jones, M.R.; Mirkin, C.A. Transmutable nanoparticles with reconfigurable surface ligands. *Science* **2016**, *351*, 579–582. [CrossRef]
28. Auyeung, E.; Li, T.I.N.G.; Senesi, A.J.; Schmucker, A.L.; Pals, B.C.; de la Cruz, M.O.; Mirkin, C.A. DNA-mediated nanoparticle crystallization into Wulff polyhedra. *Nature* **2014**, *505*, 73–77. [CrossRef]
29. Jishkariani, D.; Diroll, B.T.; Cargnello, M.; Klein, D.R.; Hough, L.A.; Murray, C.B.; Donnio, B. Dendron-mediated engineering of interparticle separation and self-assembly in dendronized gold nanoparticles superlattices. *J. Am. Chem. Soc.* **2015**, *137*, 10728–10734. [CrossRef]
30. Jishkariani, D.; Lee, J.D.; Yun, H.; Paik, T.; Kikkawa, J.M.; Kagan, C.R.; Donnio, B.; Murray, C.B. The dendritic effect and magnetic permeability in dendron coated nickel and manganese zinc ferrite nanoparticles. *Nanoscale* **2017**, *9*, 13922–13928. [CrossRef]
31. Ye, X.; Chen, J.; Murray, C.B. Polymorphism in self-assembled AB6 binary nanocrystal superlattices. *J. Am. Chem. Soc.* **2011**, *133*, 2613–2620. [CrossRef] [PubMed]
32. Kang, Y.; Ye, X.; Chen, J.; Qi, L.; Diaz, R.E.; Doan-nguyen, V.; Xing, G.; Kagan, C.R.; Li, J.; Gorte, R.J.; et al. Engineering catalytic contacts and thermal stability: Gold/iron oxide binary nanocrystal superlattices for CO oxidation. *J. Am. Chem. Soc.* **2013**, *135*, 1499–1505. [CrossRef] [PubMed]
33. Redl, F.X.; Cho, K.; Murray, C.B.; Brien, S.O. Three-dimensional binary superlattices of magnetic nanocrystals and semiconductor quantum dots. *Nautre* **2003**, *423*, 968–971. [CrossRef] [PubMed]
34. Dai, Z.R.; Sun, S.; Wang, Z.L. Phase transformation, coalescence, and twinning of monodisperse FePt nanocrystals. *Nano Lett.* **2001**, *1*, 443–447. [CrossRef]
35. Dong, A.; Chen, J.; Ye, X.; Kikkawa, J.M.; Murray, C.B. Enhanced thermal stability and magnetic properties in NaCl-type. *J. Am. Chem. Soc.* **2011**, *133*, 13296–13299. [CrossRef] [PubMed]
36. Dong, A.; Ye, X.; Chen, J.; Murray, C.B. Two-dimensional binary and ternary nanocrystal superlattices: The case of monolayers and bilayers. *Nano Lett.* **2011**, *11*, 1804–1809. [CrossRef]
37. Paik, T.; Yun, H.; Fleury, B.; Hong, S.H.; Jo, P.S.; Wu, Y.; Oh, S.J.; Cargnello, M.; Yang, H.; Murray, C.B.; et al. Hierarchical materials design by pattern transfer printing of self-assembled binary nanocrystal superlattices. *Nano Lett.* **2017**, *17*, 1387–1394. [CrossRef]
38. Murray, M.J.; Sanders, J.V. Close-packed structures of spheres of two different sizes II. The packing densities of likely arrangements. *Philos. Mag. A Phys. Condens. Matter Struct. Defects Mech. Prop.* **1980**, *42*, 721–740. [CrossRef]
39. Talapin, D.V.; Shevchenko, E.V.; Bodnarchuk, M.I.; Ye, X.; Chen, J.; Murray, C.B. Quasicrystalline order in self-assembled binary nanoparticle superlattices. *Nature* **2009**, *461*, 964–967. [CrossRef]
40. Evers, W.H.; Friedrich, H.; Filion, L.; Dijkstra, M.; Vanmaekelbergh, D. Observation of a ternary nanocrystal superlattice and its structural characterization by electron tomography. *Angew. Chem. Int. Ed.* **2009**, *48*, 9655–9657. [CrossRef]
41. Wang, P.P.; Qiao, Q.; Zhu, Y.; Ouyang, M. Colloidal binary supracrystals with tunable structural lattices. *J. Am. Chem. Soc.* **2018**, *140*, 9095–9098. [CrossRef] [PubMed]
42. Yang, Y.; Wang, B.; Shen, X.; Yao, L.; Wang, L.; Chen, X.; Xie, S.; Li, T.; Hu, J.; Yang, D.; et al. Scalable assembly of crystalline binary nanocrystal superparticles and their enhanced magnetic and electrochemical properties. *J. Am. Chem. Soc.* **2018**, *140*, 15038–15047. [CrossRef] [PubMed]
43. Ye, X.; Jin, L.; Caglayan, H.; Chen, J.; Xing, G.; Zheng, C. Improved size-tunable synthesis of monodisperse gold nanorods through the use of aromatic additives. *ACS Nano* **2012**, *6*, 2804–2817. [CrossRef] [PubMed]
44. Noh, S.-H.; Na, W.; Jang, J.-T.; Lee, J.-H.; Lee, E.J.; Moon, S.H.; Lim, Y.; Shin, J.-S.; Cheon, J. Nanoscale magnetism control via surface and exchange anisotropy for optimized ferrimagnetic hysteresis. *Nano Lett.* **2012**, *12*, 3716–3721. [CrossRef] [PubMed]

45. Gordon, T.R.; Cargnello, M.; Paik, T.; Mangolini, F.; Weber, R.T.; Fornasiero, P.; Murray, C.B. Nonaqueous synthesis of TiO2 nanocrystals using TiF4 to engineer morphology, oxygen vacancy concentration, and photocatalytic activity. *J. Am. Chem. Soc.* **2012**, *134*, 6751–6761. [CrossRef] [PubMed]
46. Huynh, W.U.; Dittmer, J.J.; Alivisatos, A.P. Hybrid nanorod-polymer solar cells. *Science* **2002**, *295*, 2425–2427. [CrossRef] [PubMed]
47. Talapin, D.V.; Shevchenko, E.V.; Murray, C.B.; Kornowski, A.; Föraster, S.; Weller, H. CdSe and CdSe/CdS nanorod solids. *J. Am. Chem. Soc.* **2004**, *126*, 12984–12988. [CrossRef] [PubMed]
48. Paik, T.; Gordon, T.R.; Prantner, A.M.; Yun, H.; Murray, C.B. Designing tripodal and triangular gadolinium oxide nanoplates and self-assembled nanofibrils as potential multimodal bioimaging probes. *ACS Nano* **2013**, *7*, 2850–2859. [CrossRef] [PubMed]
49. Paik, T.; Ko, D.K.; Gordon, T.R.; Doan-Nguyen, V.; Murray, C.B. Studies of liquid crystalline self-assembly of GdF3 nanoplates by in-plane, out-of-plane SAXS. *ACS Nano* **2011**, *5*, 8322–8330. [CrossRef] [PubMed]
50. Sigman, M.B.; Ghezelbash, A.; Hanrath, T.; Saunders, A.E.; Lee, F.; Korgel, B.A. Solventless synthesis of monodisperse Cu2S nanorods, nanodisks, and nanoplatelets. *J. Am. Chem. Soc.* **2003**, *125*, 16050–16057. [CrossRef]
51. Chen, M.; Pica, T.; Jiang, Y.; Li, P.; Yano, K.; Liu, J.P.; Datye, A.K.; Fan, H. Synthesis and self-assembly of fcc phase FePt nanorods. *J. Am. Chem. Soc.* **2007**, *129*, 6348–6349. [CrossRef] [PubMed]
52. Singh, G.; Chan, H.; Baskin, A.; Gelman, E.; Repnin, N.; Král, P.; Klajn, R. Self-assembly of magnetite nanocubes into helical superstructures. *Science* **2014**, *345*, 1149–1153. [CrossRef] [PubMed]
53. Ye, X.; Millan, J.A.; Engel, M.; Chen, J.; Diroll, B.T.; Glotzer, S.C.; Murray, C.B. Shape alloys of nanorods and nanospheres from self-assembly. *Nano Lett.* **2013**, *13*, 4980–4988. [CrossRef] [PubMed]
54. Paik, T.; Diroll, B.T.; Kagan, C.R.; Murray, C.B. Binary and ternary superlattices self-assembled from colloidal nanodisks and nanorods. *J. Am. Chem. Soc.* **2015**, *137*, 6662–6669. [CrossRef] [PubMed]
55. Paik, T.; Murray, C.B. Shape-directed binary assembly of anisotropic nanoplates: A nanocrystal puzzle with shape-complementary building blocks. *Nano Lett.* **2013**, *13*, 2952–2956. [CrossRef] [PubMed]
56. Chen, J.; Dong, A.; Cai, J.; Ye, X.; Kang, Y.; Kikkawa, J.M.; Murray, C.B. Collective dipolar interactions in self-assembled magnetic binary nanocrystal superlattice membranes. *Nano Lett.* **2010**, *10*, 5103–5108. [CrossRef] [PubMed]
57. Urban, J.J.; Talapin, D.V.; Shevchenko, E.V.; Kagan, C.R.; Murray, C.B. Synergism in binary nanocrystal superlattices leads to enhanced p-type conductivity in self-assembled PbTe/Ag2Te thin films. *Nat. Mater.* **2007**, *6*, 115–121. [CrossRef] [PubMed]
58. Song, J.H.; Jeong, S. Colloidal quantum dot based solar cells: From materials to devices. *Nano Converg.* **2017**, *4*, 21. [CrossRef] [PubMed]
59. Shevchenko, E.V.; Ringler, M.; Schwemer, A.; Talapin, D.V.; Klar, T.A.; Rogach, A.L.; Feldmann, J.; Alivisatos, A.P. Self-assembled binary superlattices of CdSe and Au nanocrystals and their fluorescence properties. *J. Am. Chem. Soc.* **2008**, *130*, 3274–3275. [CrossRef] [PubMed]
60. Gaulding, E.A.; Diroll, B.T.; Goodwin, E.D.; Vrtis, Z.J.; Kagan, C.R.; Murray, C.B. Deposition of wafer-scale single-component and binary nanocrystal superlattice thin films via dip-coating. *Adv. Mater.* **2015**, *27*, 2846–2851. [CrossRef] [PubMed]

 © 2019 by the authors. Licensee MDPI, Basel, Switzerland. This article is an open access article distributed under the terms and conditions of the Creative Commons Attribution (CC BY) license (http://creativecommons.org/licenses/by/4.0/).

Article

Preparation of CuCrO$_2$ Hollow Nanotubes from an Electrospun Al$_2$O$_3$ Template

Hsin-Jung Wu [1,†], Yu-Jui Fan [2,†], Sheng-Siang Wang [1], Subramanian Sakthinathan [1], Te-Wei Chiu [1,*], Shao-Sian Li [1,3,*] and Joon-Hyeong Park [4]

1. Department of Materials and Mineral Resources Engineering, National Taipei University of Technology, 1, Sec. 3, Zhongxiao E. Rd., Taipei 10608, Taiwan
2. School of Biomedical Engineering, Taipei Medical University, No. 250, Wuxsing Street, Taipei 11031, Taiwan
3. Graduate Institute of Biomedical Optomechatronics, College of Biomedical Engineering, Taipei Medical University, No. 250, Wuxing Street, Taipei 11031, Taiwan
4. Birck Nanotechnology Center, Purdue University, West Lafayette, IN 47907, USA
* Correspondence: tewei@ntut.edu.tw (T.-W.C.); ssli@tmu.edu.tw (S.-S.L.); Tel.: +886-02-2771-2171 (T.-W.C.); +886-02-6638-2736 (S.-S.L.)
† These authors contributed equally to this work.

Received: 17 July 2019; Accepted: 1 September 2019; Published: 3 September 2019

Abstract: A hollow nanostructure is attractive and important in different fields of applications, for instance, solar cells, sensors, supercapacitors, electronics, and biomedical, due to their unique structure, large available interior space, low bulk density, and stable physicochemical properties. Hence, the need to prepare hollow nanotubes is more important. In this present study, we have prepared CuCrO$_2$ hollow nanotubes by simple approach. The CuCrO$_2$ hollow nanotubes were prepared by applying electrospun Al$_2$O$_3$ fibers as a template for the first time. Copper chromium ions were dip-coated on the surface of electrospun-derived Al$_2$O$_3$ fibers and annealed at 600 °C in vacuum to form Al$_2$O$_3$-CuCrO$_2$ core-shell nanofibers. The CuCrO$_2$ hollow nanotubes were obtained by removing Al$_2$O$_3$ cores by sulfuric acid wet etching while preserving the rest of original structures. The structures of the CuCrO$_2$-coated Al$_2$O$_3$ core-shell nanofibers and CuCrO$_2$ hollow nanotubes were identified side-by-side by X-ray diffraction, field emission scanning electron microscopy, and transmission electron microscopy. The CuCrO$_2$ hollow nanotubes may find applications in electrochemistry, catalysis, and biomedical application. This hollow nanotube preparation method could be extended to the preparation of other hollow nanotubes, fibers, and spheres.

Keywords: electrospinning; CuCrO$_2$; hollow nanotube; Al$_2$O$_3$ template; one-dimensional structures

1. Introduction

One-dimensional (1D) nanostructure materials such as nanotubes, nanobelts, and nanofibers have attracted wide interest in nanoscience and technology [1]. Regulating the size and shape of synthesized nanomaterials is of great technological interest nowadays. Particularly, hollow nanostructures have received considerable attention due to their high surface areas and structural uniqueness, thus they have been extensively applied in many fields, such as sensors, dye-sensitized solar cells, catalysts, supercapacitors, photoelectrochemical cells, electronics, and biomolecule devices. Hence, different approaches have been used in the development of hollow nanotubes and nanofibers for large-scale synthesis [2,3]. One of such structural approaches is electrospinning which has been widely applied to synthesize nanofibers from a variety of oxide materials [4].

Electrospinning is a fiber formation method that uses self-repulsion effect, which induces an electrostatic charge on a precursor material to stretch the liquid in an electric field into fiber structure. The dimension of fiber diameter ranges from tens nanometer to few micrometers [5]. In the past few years, it has been an effective method to prepare polymer-based nano- or microfibers. Different kinds of polymers have been successfully electrospun from melts or solutions into ultrathin fibers [6]. Up to date, the preparation of nanofibers with solid cross-sections has been studied [7,8].

P-type transparent conducting oxides with delafossite structure has been demonstrated with potential applications in various fields including organic photovoltaic (OPV) devices [9], perovskite solar cells [10], antibacterial surface [11], gas sensors [12], solid propellants [13], etc. The delafossite structure of copper-based catalysts also has great importance in catalytic steam reforming of methanol to hydrogen production and heterogeneous catalysis for chlorine production due to their high thermal stability, fine porous structure, high surface area, high selectivity, and excellent activity at low temperature. Besides, copper delafossite materials are more stable than Ru, Pd, Au, and Pt catalyst at the steam reforming process [14–16]. Cu-based delafossites have been reported including $CuAlO_2$ [17], $CuFeO_2$ [18], $CuGaO_2$ [19], $CuInO_2$ [20], $CuScO_2$ [21], $CuCrO_2$ [22], and Mg-doped $CuCrO_2$ [23,24]. The chemical formula of delafossite structure is that of a ternary oxide $A^+B^{+3}O_2$. According to the report, the delafossite structure of $CuCrO_2$ has a wide bandgap of 3.1 eV and highest conductivity among all types of semiconductors [25]. Hence, $CuCrO_2$ and $CuAlO_2$ have drawn considerable attention in optoelectronic devices [26,27]. The delafossite material consists of two alternating sheets: a planar layer of triangular-patterned cations (A) and a layer of edge-sharing BO_6 octahedrons flattened with respect to the c-axis. Depending on the orientation of layer stacking, two polytypes of delafossite oxide can be created. Considering the morphological effects, catalyst with hollow tube structure shows very promising potential because of the highly selective catalytic reaction. For example, ZSM-5/SiO_2 hollow structure catalyst selectively increases the paraxylene from the 24% to 89.6% in xylene in methanol-to-aromatics conversion [28]. A single-wall carbon nanotube/iron tetraphenyl porphyrin composite sensor shows a selectively high response toward xylene among benzene and toluene [29]. Carbon nanotube pores (CNTP) show potential to be used as next-generation water purification technologies because CNTP provides high selectivity of water and anions [30]. Further, a porous hollow tube CeO_2/Au@SiO_2 nanocatalyst exhibited excellent catalytic activity toward 4-nitrophenol reduction [31]. Platinum (Pt) functionalized NiO hollow tube exhibited remarkable selectivity of C_2H_5OH sensing against CO and H_2 gases [32]. The hollow structure of CuO@SiO_2 exhibits excellent catalytic activities toward CO and NO oxidation compared with individual CuO and SiO_2 [33]. Besides, carbon nanotube catalyst could raise the selectivity of H_2 production rather than CO [34].

However, nanotube with hollow cross-sections are challenging to fabricate because of multi-step treatments (e.g., a template process) or specially designed instrumentation facilities (e.g., for co-electrospinning with coaxial capillaries) [35]. Nanofiber (7.85 m²/g) [36] or nanopowder structures (30.92 m²/g) [37], such as hollow nanotubes (136 m²/g), have a higher surface-to-volume ratio and higher porosity, which are favorable for adsorption in catalysis [38]. Hence, developing a simple approach to obtain hollow nanotubes is of great importance. [36,39]. In this study, the main objective was to explore the use of Al_2O_3 microfibers as a template to prepare a core-shell structure of Al_2O_3-$CuCrO_2$ by immersion in Cu-Cr-O precursor solution. The alumina structure was then removed by etching in H_2SO_4 to form the $CuCrO_2$ hollow nanotubes.

2. Materials and Methods

All the high-purity chemicals used in this experiment were obtained from Sigma Chemical Co, Taiwan. The electrospun Al_2O_3 microfibers precursor was prepared by the electrospinning method. Typically, the precursor solution was prepared by dissolving aluminum nitrate ($Al(NO_3)_3 \cdot 9H_2O$) into 14.4 mL of dimethylformamide (DMF) solvent to make a 0.04 M metal source solution. Then, 2.4 g polyvinylpyrrolidone (Mw = 1,300,000) was mixed into the aforementioned prepared metal source solution followed by constant stirring for 6 h. Finally, a viscous gel-like precursor solution of Al_2O_3 was obtained. The Al_2O_3 precursor solution was loaded into a horizontal programmable syringe pump. A schematic image of the fundamental electrospinning process is illustrated in Figure 1. An ordinary electrospinning set-up, a high-voltage source is combined with the metallic needle, which is connected to a syringe pump. This syringe pump was connected with Teflon tube (length = 125 mm, diameter = 4.2 mm) for conventional electrospinning setup. During the electrospinning process, the precursor solution was placed in a 10 mL syringe equipped with a stainless steel needle (ID = 0.5 mm). A voltage of 20 kV was applied to the stainless steel needle tip, and the collector was fixed at a distance of 16 cm from the needle tip with the flow controlled at 0.02 mL/h. The electrospun Al_2O_3 precursor was distributed uniformly over the collector to form Al_2O_3 precursor fibers (Step 1). After the electrospinning, the electrospun Al_2O_3 precursor fibers were heated at a rate of 5 °C/min to the annealing temperature of 600 °C in a high-temperature furnace at air atmosphere and then held at that temperature for 2 h, after which Al_2O_3 nanofibers were formed (Step 2) and the diameter of the Al_2O_3 nanofibers is <100 nm.

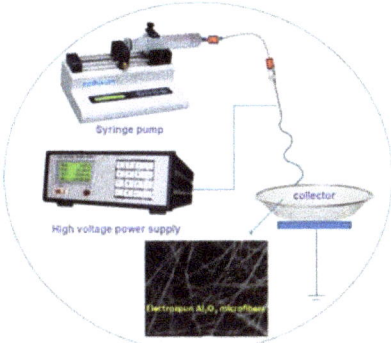

Figure 1. Schematic illustration of electrospinning preparation of as-spun fiber.

2.1. Preparation of $CuCrO_2$ Hollow Nanotube

Copper (II) acetate, chromium (III) acetate, and ethanolamine were dissolved in ethylene glycol monomethyl ether (30 mL) to obtain 0.2 M precursor. The prepared solution was stirred for 24 h to obtain a well-mixed solution without impurities. Al_2O_3 microfibers were dipped in Cu-Cr-O ion solution up to 3 sec to deposit Cu-Cr-O ions on the fiber surfaces and form an Al_2O_3-Cu-Cr-O core (Step 3). The Cu-Cr-O ions deposited on Al_2O_3 fibers were dried at 80 °C on a hotplate for 2 min. Then the coated fibers were annealed at 600 °C in vacuum (Step 4). After that, the prepared nanofibers were etched with 2 M H_2SO_4 to remove the Al_2O_3 and other minor impurities from the fibers (Step 5) [39]. The nanofibers were repeatedly rinsed with DI water and a centrifuge was used to separate the liquid and fibers. Finally, the collected nanofibers were dried in an oven at 80 °C to form $CuCrO_2$ hollow nanotube (Figure 2).

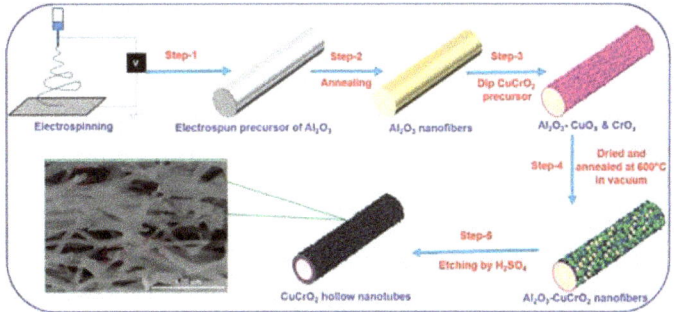

Figure 2. Schematic illustration of $CuCrO_2$ hollow nanotubes fabrication process.

2.2. Characterization

The crystallized phase of Al_2O_3 microfibers and $CuCrO_2$ hollow nanotubes was characterized with an X-ray diffractometer (XRD, D_2 Phaser, Bruker) with Cu Kα radiation (λ = 0.15418 nm) from 20° to 80°, a working voltage of 30 kV, and current of 10 mA. The thermal decomposition behavior of the as-spun fibers was identified using a thermogravimetric analysis/differential scanning calorimeter (TGA/DSC, STA 449 F5, NETZSCH) at a heating rate of 10 °C/min. The surface morphology and structure of the nanofibers were observed by field emission scanning electron microscopy (FE-SEM, Hitachi S-4700) SEM 15 kV, 10 cm SEI detector, and nanotubes were identified by transmission electron microscopy (TEM, JEM-2100F, JEOL) operated at a working voltage of 200 kV, working current was 10 µA and chamber was about 1.0×10^{-6} to 3.0×10^{-6} torr. The composition hollow nanotubes were confirmed by JOEL JEM2100F type scanning transmission electron microscope (STEM) attached with an energy dispersive spectrometer (EDS).

3. Results

3.1. TGA Analysis

The TGA/DSC analysis of the Al_2O_3 electrospun fibers studied at a heating rate of 10 °C/min in air is shown in Figure 3. Two discrete regions of electrospun fibers weight loss occurred at about 135 °C and 300 °C. The weight loss at around 135 °C could be attributed to DMF solvent. Exothermic peaks at 300 °C with a large weight loss of ~80% corresponded to the decomposition of nitrate, PVP polymer, and other minor organic constituents during the burning combustion. For temperature higher than 600 °C, there was almost no change in the TGA curve, which confirmed that the complete decomposition of organic materials and polymer during the formation of Al_2O_3 fibers [40–43].

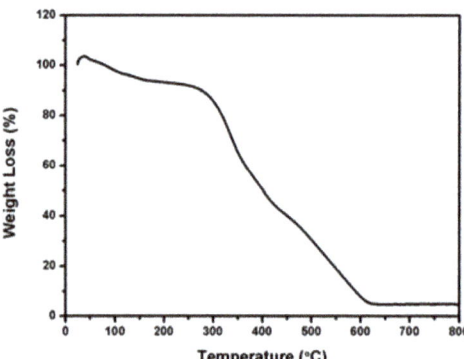

Figure 3. Thermogravimetric-derivative thermal analysis of as-spun Al_2O_3 precursor microfibers recorded in air at a heating rate of 10 °C/min.

3.2. X-ray Diffraction Investigation

Figure 4 shows the XRD analysis of annealed Al_2O_3 fibers prepared by electrospinning method. The Al_2O_3 fibers were fabricated following the process mentioned in the last section with thermal annealing at elevated temperature for 2 h. We found no distinct diffraction peak for the as-spun fibers, but after the fibers were annealed at 600 °C, a clear amorphous phase was found. The XRD pattern indicated that the Al_2O_3 fibers became crystallized when the annealing temperature was over 800 °C [44].

Figure 5 shows the XRD pattern of Al_2O_3 fibers with copper chromium ions deposited on the surfaces after annealing in vacuum at 600 °C for 30 min and 60 min, and at 700 °C for 30 min. The fibers were composed of an Al_2O_3 core and the copper chromium ion solution. The XRD studies show the peaks of Al_2O_3 for the fibers annealed at 600 °C for 60 min. It is presumed that the prolonged annealing time caused the crystallization of alumina [39,44].

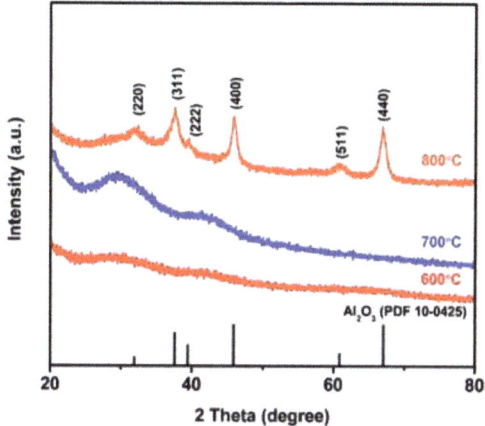

Figure 4. XRD patterns of electrospun Al_2O_3 precursor fibers annealed for 2 h in the air at various temperatures (600 °C–800 °C).

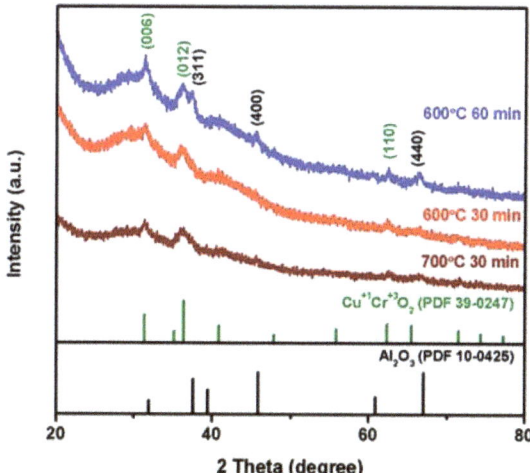

Figure 5. XRD patterns of Al_2O_3 microfibers with copper chromium oxide deposited on the surfaces after annealing at 600 °C and 700 °C in vacuum.

Figure 6 shows the XRD pattern of Al_2O_3 fibers with copper chromium ion solution deposited on the surfaces after annealing at 600 °C for 30 min in vacuum followed by leaching with 2M H_2SO_4 solution due to the strong acid and without the formation of impurities. That solution was employed because Al_2O_3 is an amphoteric oxide and reacts with both acid and alkaline solutions. From comparing Figure 6 with Figure 5, it is clear that the main phase of $CuCrO_2$ can be clearly seen in the XRD pattern after the acid immersion. For comparison, NaOH solution was also used to remove alumina cores. As can be seen from the figures, after immersion of the fibers in NaOH solution, only the CuO phase remain while the chromium oxide disappeared. Therefore, we concluded that Al_2O_3 fibers with copper chromium ion solution deposited on the surfaces could be treated with 2M H_2SO_4 solution and DI water to obtain $CuCrO_2$ hollow nanotube [39].

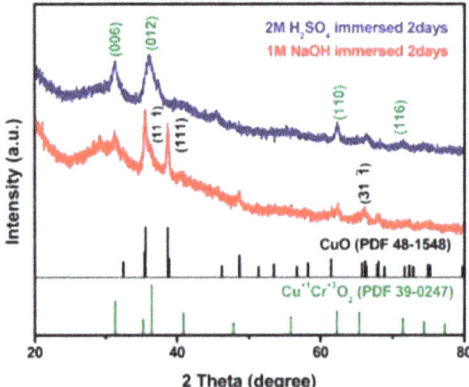

Figure 6. XRD patterns of Al_2O_3 microfibers with copper chromium oxide deposited on the surfaces after annealing at 600 °C in vacuum followed by leaching with 2M H_2SO_4 and NaOH solution.

3.3. SEM Analysis

The SEM micrographs of as-spun Al_2O_3 precursor fibers have fine cylindrical with smooth surface morphology and shows in Scheme 1 [41]. Besides, the SEM image of Al_2O_3 electrospun fibers annealed for 2 h in air at 600 °C and 800 °C are presented in Figure 7. The morphology of the fibers reveals that the Al_2O_3 fibers have continuous, one-dimensional structure and that the diameter of each Al_2O_3 fiber is <100 nm. The morphology and dimension of Al_2O_3 fibers are essentially similar in the case of annealing temperature of 600 °C and the counterpart in 800 °C.

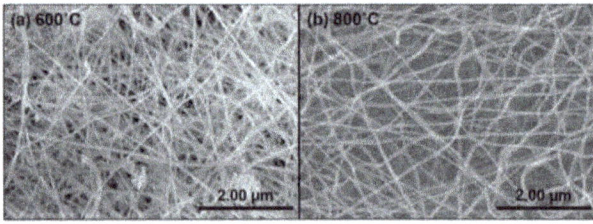

Figure 7. SEM images of electrospun Al_2O_3 microfibers annealed for 2 h at (**a**) 600 °C and (**b**) 800 °C.

Figure 8 shows the morphology of Al_2O_3 fibers immersed in copper chromium ion solution and then dried for 2 min at 80 °C on a hotplate. After that, the Al_2O_3-$CuCrO_2$ fibers were annealed in vacuum at 600 °C for 30 min (Figure 8a) and 60 min (Figure 8b), and at 700 °C for 30 min (Figure 8c). The surfaces of the fibers are smooth, and there is no specific change compared with calcined amorphous Al_2O_3 fibers. The copper chromium ion precursor solution, composed of mixed copper acetate, chromium acetate, and ethanolamine, was dissolved in ethylene glycol monomethyl ether.

Figure 9 shows a SEM image of Al_2O_3-$CuCrO_2$ nanofibers after immersion in 2M H_2SO_4 and oven-drying at 80 °C for 1 day. As can be seen from the SEM morphology, there is a hollow-like structure at the tip of the $CuCrO_2$ nanotubes etched by 2M H_2SO_4. It was inferred that the Al_2O_3 core was mostly removed by the H_2SO_4 solution and remaining impurities were removed by DI water.

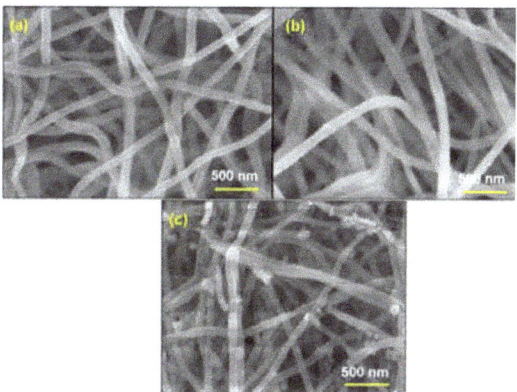

Figure 8. SEM images of Al$_2$O$_3$-CuCrO$_2$ nanofibers annealed at 600 °C for (**a**) 30 min, (**b**) 60 min, and at 700 °C for (**c**) 30 min.

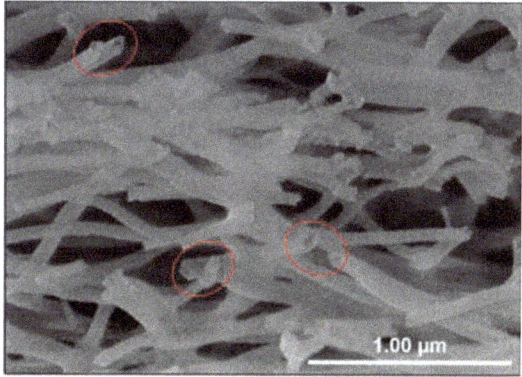

Figure 9. SEM images of CuCrO$_2$ hollow nanotubes after removal of Al$_2$O$_3$ core and impurities by 2M H$_2$SO$_4$ for 2 days and DI water.

3.4. TEM Analysis

To identify the structure of the CuCrO$_2$ hollow nanotubes synthesized by annealing and followed by chemical etching, TEM was used to further confirm the hollow structures of the nanotubes. The nanotubes were formed by using Al$_2$O$_3$ fiber as a template and depositing copper chromium ions on the tube surfaces so that the inner core was Al$_2$O$_3$. As shown in TEM image in Figure 10, the inner template of Al$_2$O$_3$ was completely etched away by 2M H$_2$SO$_4$ solution. The inner diameter of the nanotubes was about 70 nm, which is consistent with the diameter of Al$_2$O$_3$ fiber. The tube wall which consists of CuCrO$_2$ features a thickness of several tens of nanometer [39]. These results indicate that the chemical etching method was successful in making CuCrO$_2$ hollow nanotubes. Based on previous report, CuCrO$_2$ hollow nanotubes have more porous cavity than none-hollow CuCrO$_2$ nanofibers due to annealing condition [10].

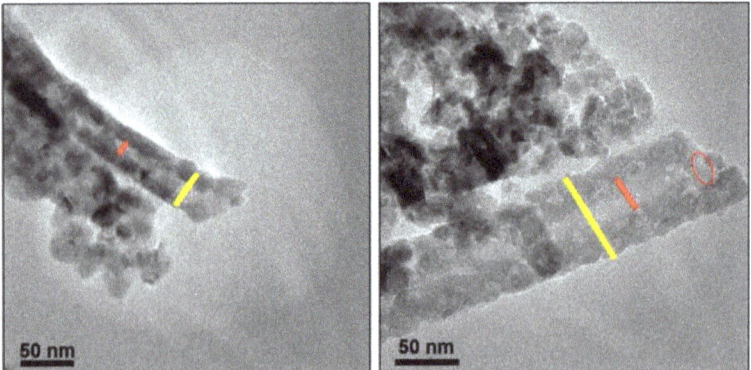

Figure 10. TEM images of the CuCrO$_2$ hollow nanotubes.

3.5. STEM Analysis

Figure 11a shows a STEM image of CuCrO$_2$ hollow nanotube formed by annealing and chemical etching. The average diameter of the CuCrO$_2$ hollow nanotube was about 100 nm and that of the center hollow was approximately 20 nm. These results exhibit that the chemical etching method succeeded in producing hollow nanotube. The STEM-EDS signals of CuCrO$_2$ nanotube showed the presence of (Figure 11b) Cu, (Figure 11c) Cr, and (Figure 11d) O. Besides, the STEM-EDS spectrum showed higher numbers of atoms present in the tube edge than inside the cavity, which clearly shows the successful formation of the CuCrO$_2$ hollow nanotubes.

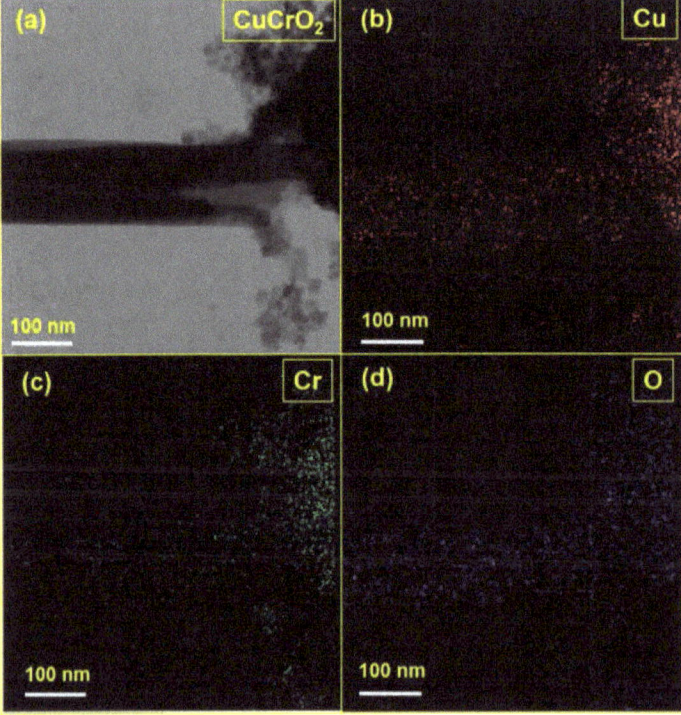

Figure 11. (a) STEM image of the CuCrO$_2$ hollow nanotube, (b) Cu, (c) Cr, (d) O.

4. Conclusions

CuCrO$_2$ hollow nanotubes were successfully prepared by our proposed method using electrospun Al$_2$O$_3$ fiber as core template. The amorphous Al$_2$O$_3$ fibers were prepared by annealing the as-spun alumina precursor fibers at 600 °C for 2 h. These continuous and one-dimensional fibers were then deposited with CuCrO$_2$ precursor and formed CuCrO$_2$ cladding layer by thermal annealing at 600 °C for 30 min. After removing amorphous Al$_2$O$_3$ core fibers by using H$_2$SO$_4$, CuCrO$_2$ nanotubes with an inner diameter of 70 nm and tube wall thickness of 30 nm were obtained. This work demonstrated a simple solution-based approach for the synthesis of oxide nanotubes and could be further extended to synthesize oxide materials with various complicated hollow structures.

Author Contributions: Conceptualization, H.-J.W. and T.-W.C.; data curation, H.-J.W. and T.-W.C.; formal analysis, H.-J.W.; funding acquisition, Y.-J.F., T.-W.C., and S.-S.L.; investigation, T.-W.C.; methodology, H.-J.W.; project administration, T.-W.C.; resources, T.-W.C.; supervision, T.-W.C.; validation, Y.-J.F., S.S., T.-W.C., and S.-S.L.; visualization, H.-J.W. and S.S.; writing—original draft, H.-J.W., S.-S.W., and T.-W.C.; writing—review and editing, H.-J.W., S.S., T.-W.C., S.-S.L., and J.-H.P.

Funding: This research was funded by University System of Taipei Joint Research Program (USTP-NTUT-TMU-108-06) and Ministry of Science and Technology, Taiwan grant number (MOST106-2221-E-027-041).

Acknowledgments: This work was supported by the Ministry of Science and Technology of Taiwan (MOST106-2221-E-027-041) and University System of Taipei Joint Research Program (USTP-NTUT-TMU-108-06). The authors appreciate the Precision Research and Analysis Center of National Taipei University of Technology (NTUT) for providing the measurement facilities.

Conflicts of Interest: The authors declare no conflict of interest.

References

1. Huang, J.Y.; Zhong, L.; Wang, C.M.; Sullivan, J.P.; Xu, W.; Zhang, L.Q.; Mao, S.X.; Hudak, N.S.; Liu, X.H.; Subramanian, A.; et al. In situ observation of the electrochemical lithiation of a single SnO$_2$ nanowire electrode. *Science* **2010**, *330*, 1515–1521. [CrossRef] [PubMed]
2. Qian, H.; Lin, G.; Zhang, Y.; Gunawan, P.; Xu, R. A new approach to synthesize uniform metal oxide hollow nanospheres via controlled precipitation. *Nanotechnology* **2007**, *18*, 355602–355608. [CrossRef]
3. Zhan, G.; Zeng, H.C. General strategy for preparation of carbon-nanotube-supported nanocatalysts with hollow cavities and mesoporous shells. *Chem. Mater.* **2015**, *27*, 726–734. [CrossRef]
4. Choi, S.W.; Park, J.Y.; Kim, S.S. Growth behavior and sensing properties of nanograins in CuO nanofibers. *Chem. Eng. J.* **2011**, *172*, 550–556.
5. Chiu, T.W.; Tu, C.H.; Chen, Y.T. Fabrication of electrospun CuCr$_2$O$_4$ fibers. *Ceram. Int.* **2015**, *41*, S399–S406. [CrossRef]
6. Inagaki, M.; Yang, Y.; Kang, F. Carbon nanofibers prepared via electrospinning. *Adv. Mater.* **2012**, *24*, 2547–2566. [CrossRef] [PubMed]
7. Acik, M.; Baristiran, C.; Sonmez, G. Highly surfaced polypyrrole nano-networks and nano-fibers. *J. Mater. Sci.* **2006**, *41*, 4678–4683. [CrossRef]
8. Zhao, M.; Wang, X.; Ning, L.; He, H.; Jia, J.; Zhang, L.; Li, X. Synthesis and optical properties of Mg-doped ZnO nanofibers prepared by electrospinning. *J. Alloys Compd.* **2010**, *507*, 97–100. [CrossRef]
9. Wang, J.; Daunis, T.B.; Cheng, L.; Zhang, B.; Kim, J.; Hsu, J.W.P. Combustion synthesis of p-type transparent conducting CuCrO$_{2+x}$ and Cu: CrO$_x$ thin films at 180 °C. *ACS Appl. Mater. Interfaces* **2018**, *10*, 3732–3738. [CrossRef]
10. Shohl, W.A.D.; Daunis, T.B.; Wang, X.; Wang, J.; Zhang, B.; Barrera, D.; Yan, Y.; Hsu, J.W.P.; Mitzi, D.B. Room-temperature fabrication of a delafossite CuCrO$_2$ hole transport layer for perovskite solar cells. *J. Mater. Chem. A* **2018**, *6*, 469–477. [CrossRef]
11. Chiu, T.W.; Yang, Y.C.; Yeh, A.C.; Wang, Y.P.; Feng, Y.W. Antibacterial property of CuCrO$_2$ thin films prepared by RF magnetron sputtering deposition. *Vacuum* **2013**, *87*, 174–177. [CrossRef]
12. Tong, B.; Deng, Z.; Xu, B.; Meng, G.; Shao, J.; Liu, H.; Dai, T.; Shan, X.; Dong, W.; Wang, S.; et al. Oxygen vacancy defects boosted high performance p-Type delafossite CuCrO$_2$ gas sensors. *ACS Appl. Mater. Interfaces* **2018**, *10*, 34727–34734. [CrossRef] [PubMed]

13. Sathiskumar, P.S.; Thomas, C.R.; Madras, G. Solution combustion synthesis of nanosized copper chromite and its use as a burn rate modifier in solid propellants. *Ind. Eng. Chem. Res.* **2012**, *51*, 10108–10116. [CrossRef]
14. Chiu, T.W.; Hong, R.T.; Yu, B.S.; Huang, Y.H.; Kameoka, S.; Tsai, A.P. Improving steam-reforming performance by nanopowdering $CuCrO_2$. *Int. J. Hydrogen Energy* **2014**, *39*, 14222–14226. [CrossRef]
15. Hwang, B.Y.; Sakthinathan, S.; Chiu, T.W.; Chuang, C.H.; Fu, Y.; Yu, B.S. Production of hydrogen from steam reforming of methanol carried out by self-combusted $CuCr_{1-x}Fe_xO_2$ (x = 0–1) nanopowders catalyst. *Int. J. Hydrogen Energy* **2019**, *44*, 2848–2856. [CrossRef]
16. Amrute, A.P.; Larrazabal, G.O.; Mondelli, C.; Ramirez, J.P. $CuCrO_2$ delafossite: A stable copper catalyst for chlorine production. *Angew. Chem. Int. Ed.* **2013**, *52*, 9772–9775. [CrossRef]
17. Fang, M.; He, H.; Lu, B.; Zhang, W.; Zhao, B.; Ye, Z.; Huang, J. Optical properties of p-type $CuAlO_2$ thin film grown by rf magnetron sputtering. *Appl. Surf. Sci.* **2011**, *257*, 8330–8333. [CrossRef]
18. Chen, H.; Wu, J. Transparent conductive $CuFeO_2$ thin films prepared by sol gel processing. *Appl. Surf. Sci.* **2012**, *258*, 4844–4847. [CrossRef]
19. Ueda, K.; Hase, T.; Yanagi, H.; Kawazoe, H.; Hosono, H.; Ohta, H.; Orita, M.; Hirano, M. Epitaxial growth of transparent p-type conducting $CuGaO_2$ thin films on sapphire (001) substrates by pulsed laser deposition. *J. Appl. Phys.* **2001**, *89*, 1790–1793. [CrossRef]
20. Lee, J.; Heo, Y.; Lee, J.; Kim, J. Growth of $CuInO_2$ thin film using highly dense Cu_2O-In_2O_3 composite targets. *Thin Solid Films* **2009**, *518*, 1234–1237. [CrossRef]
21. Liu, F.; Makino, T.; Hiraga, H.; Fukumura, T.; Kong, Y.; Kawasaki, M. Ultrafast dynamics of excitons in delafossite $CuScO_2$ thin films. *Appl. Phys. Lett.* **2010**, *96*, 211904. [CrossRef]
22. Lim, W.T.; Sadik, P.W.; Norton, D.P.; Pearton, S.J.; Ren, F. Dry etching of $CuCrO_2$ thin films. *Appl. Surf. Sci.* **2008**, *254*, 2359–2363. [CrossRef]
23. Chiu, T.W.; Shih, J.H.; Chang, C.H. Preparation and properties of $CuCr_{1-x}Fe_xO_2$ thin films prepared by chemical solution deposition with two-step annealing. *Thin Solid Films* **2016**, *618*, 151–158. [CrossRef]
24. Wang, Y.; Gu, Y.; Wang, T.; Shi, W. Structural, optical and electrical properties of Mg-doped $CuCrO_2$ thin films by sol gel processing. *J. Alloys Compd.* **2011**, *509*, 5897–5902. [CrossRef]
25. Nagarajan, R.; Draeseke, A.D.; Sleight, A.W.; Tate, J. p-type conductivity in $CuCr_{1-x}Mg_xO_2$ films and powders. *J. Appl. Phys.* **2001**, *89*, 8022–8025. [CrossRef]
26. Kawazoe, H.; Yasukawa, M.; Hyodo, H.; Kurita, M.; Yanagi, H.; Hosono, H. P-type electrical conduction in transparent thin films of $CuAlO_2$. *Nature* **1997**, *389*, 939–942. [CrossRef]
27. Saadi, S.; Bouguelia, A.; Trari, M. Photocatalytic hydrogen evolution over $CuCrO_2$. *Sol. Energy* **2006**, *80*, 272–280. [CrossRef]
28. Zhang, J.; Qian, W.; Kong, C.; Wei, F. Increasing para-Xylene selectivity in making aromatics from methanol with a surface-modified Zn/P/ZSM-5 Catalyst. *ACS Catal.* **2015**, *5*, 2982–2988. [CrossRef]
29. Rushi, A.D.; Datta, K.P.; Ghosh, P.S.; Mulchandani, A.; Shirsat, M.D. Selective discrimination among benzene, toluene, and xylene: Probing metalloporphyrin-functionalized single-walled carbon nanotube-based field effect transistors. *J. Phys. Chem. C* **2014**, *118*, 24034–24041. [CrossRef]
30. Tunuguntla, R.H.; Henley, R.Y.; Yao, Y.C.; Pham, T.A.; Wanunu, M.; Noy, A. Enhanced water permeability and tunable ion selectivity in subnanometer carbon nanotube porins. *Science* **2017**, *357*, 792–796. [CrossRef]
31. Zhang, Z.; Shi, H.; Wu, Q.; Bu, X.; Yang, Y.; Zhang, J.; Huang, Y. MOF-derived CeO_2/Au@SiO_2 hollow nanotubes and their catalytic activity toward 4-nitrophenol reduction. *New J. Chem.* **2019**, *43*, 4581–4589. [CrossRef]
32. Cho, N.G.; Woo, H.S.; Lee, J.H.; Kim, I.D. Thin-walled NiO tubes functionalized with catalytic Pt for highly selective C_2H_5OH sensors using electrospun fibers as a sacrificial template. *Chem. Commun.* **2011**, *47*, 11300–11302. [CrossRef] [PubMed]
33. Niu, X.; Zhao, T.; Yuan, F.; Zhu, Y. Preparation of hollow CuO@SiO_2 spheres and its catalytic performances for the NO+CO and CO oxidation. *Sci. Rep.* **2015**, *5*, 9153. [CrossRef] [PubMed]
34. Zhang, Y.; Williams, P.T. Carbon nanotubes and hydrogen production from the pyrolysis catalysis or catalytic-steam reforming of waste tyres. *J. Anal. Appl. Pyrolysis* **2016**, *122*, 490–501. [CrossRef]
35. Li, P.; Shang, Z.; Cui, K.; Zhang, H.; Qiao, Z.; Zhu, C.; Zhao, N.; Xu, J. Coaxial electrospinning core-shell fibers for self-healing scratch on coatings. *Chin. Chem. Lett.* **2019**, *30*, 157–159. [CrossRef]
36. Chiu, T.W.; Yu, B.S.; Wang, Y.R.; Chen, K.T.; Lin, Y.T. Synthesis of nanosized $CuCrO_2$ porous powders via a self-combustion glycine nitrate process. *J. Alloys Compd.* **2011**, *509*, 2933–2935. [CrossRef]

37. Cetin, C.; Akyildiz, H. Production and characterization of CuCrO$_2$ nanofibers. *Mater. Chem. Phys.* **2016**, *70*, 138–144. [CrossRef]
38. Tan, Y.; Jia, Z.; Sun, J.; Wang, Y.; Cui, Z.; Guo, X. Controllable synthesis of hollow copper oxide encapsulated into N-doped carbon nanosheets as high-stability anodes for lithium-ion batteries. *J. Mater. Chem. A* **2017**, *5*, 24139–24144. [CrossRef]
39. Su, S.Y.; Wang, S.S.; Sakthinathan, S.; Chiu, T.W.; Park, J.H. Preparation of CuAl$_2$O$_4$ submicron tubes from electrospun Al$_2$O$_3$ fibers. *Ceram. Int.* **2019**, *45*, 1439–1442. [CrossRef]
40. Chiu, T.W.; Chen, Y.T. Preparation of CuCrO$_2$ nanowires by electrospinning. *Ceram. Int.* **2015**, *41*, S407–S413. [CrossRef]
41. Mahapatra, A.; Mishra, B.G.; Hota, G. Synthesis of ultra-fine α-Al$_2$O$_3$ fibers via electrospinning method. *Ceram. Int.* **2011**, *37*, 2329–2333. [CrossRef]
42. Kim, J.H.; Yoo, S.J.; Kwak, D.H.; Jung, H.J.; Kim, T.Y.; Park, K.H.; Lee, J.W. Characterization and application of electrospun alumina nanofibers. *Nanoscale Res. Lett.* **2014**, *9*, 44–49. [CrossRef] [PubMed]
43. Kang, W.; Cheng, B.; Li, Q.; Zhuang, X.; Ren, Y. A new method for preparing alumina nanofibers by electrospinning technology. *Text. Res. J.* **2011**, *81*, 148–155. [CrossRef]
44. Zhang, L.; Jiang, H.C.; Liu, C.; Dong, J.W.; Chow, P. Annealing of Al$_2$O$_3$ thin films prepared by atomic layer deposition. *J. Phys. D Appl. Phys.* **2007**, *40*, 3707–3713. [CrossRef]

© 2019 by the authors. Licensee MDPI, Basel, Switzerland. This article is an open access article distributed under the terms and conditions of the Creative Commons Attribution (CC BY) license (http://creativecommons.org/licenses/by/4.0/).

Article

Mg Doped CuCrO₂ as Efficient Hole Transport Layers for Organic and Perovskite Solar Cells

Boya Zhang [1], Sampreetha Thampy [1], Wiley A. Dunlap-Shohl [2], Weijie Xu [1], Yangzi Zheng [3], Fong-Yi Cao [4], Yen-Ju Cheng [4], Anton V. Malko [3], David B. Mitzi [2] and Julia W. P. Hsu [1,*]

[1] Department of Materials Science and Engineering, The University of Texas at Dallas, Richardson, TX 75080, USA
[2] Department of Mechanical Engineering and Materials Science, Duke University, Durham, NC 27708, USA
[3] Department of Physics, The University of Texas at Dallas, Richardson, TX 75080, USA
[4] Department of Applied Chemistry, National Chiao Tung University, 1001 University Road, Hsinchu 30010, Taiwan
* Correspondence: jwhsu@utdallas.edu

Received: 19 July 2019; Accepted: 11 September 2019; Published: 13 September 2019

Abstract: The electrical and optical properties of the hole transport layer (HTL) are critical for organic and halide perovskite solar cell (OSC and PSC, respectively) performance. In this work, we studied the effect of Mg doping on $CuCrO_2$ (CCO) nanoparticles and their performance as HTLs in OSCs and PSCs. CCO and Mg doped CCO (Mg:CCO) nanoparticles were hydrothermally synthesized. The nanoparticles were characterized by various experimental techniques to study the effect of Mg doping on structural, chemical, morphological, optical, and electronic properties of CCO. We found that Mg doping increases work function and decreases particle size. We demonstrate CCO and Mg:CCO as efficient HTLs in a variety of OSCs, including the first demonstration of a non-fullerene acceptor bulk heterojunction, and $CH_3NH_3PbI_3$ PSCs. A small improvement of average short-circuit current density with Mg doping was found in all systems.

Keywords: Mg doped $CuCrO_2$; hole transport layer; organic solar cells; perovskite solar cells

---

1. Introduction

With continued increase in power conversion efficiency (PCE), organic and perovskite solar cells (OSCs and PSCs, respectively) are promising for low cost clean electricity generation [1–3]. Further enhancing the PCE of OSCs and PSCs requires not only the development of better absorber materials, but also suitable transport layer materials. In OSCs and PSCs, the absorber is sandwiched between an electron transport layer (ETL) and a hole transport layer (HTL), whose primary functions are to set up the built-in field across the absorber and selectively extract their respective carriers, while blocking the other type of carriers. In bulk heterojunction (BHJ) OSCs, the photogenerated excitons dissociate at the donor/acceptor interface to charged carriers. In PSCs, photoabsorption directly generates electrons and holes. These carriers then drift in opposite directions due to the built-in electric field, and travel through the transport layers to the electrodes [4]. Thus, both ETL and HTL play an important role in carrier extraction and device performance. For effective hole extraction from the absorber, the material used as HTL should possess good optical and electrical properties in addition to good physical and chemical stability. Extensive studies have been devoted to organic materials, such as poly(3,4-ethylenedioxythiophene) polystyrene sulfonate and 2,2′,7,7′-Tetrakis[N,N-di(4-methoxyphenyl)amino]-9,9′-spirobifluorene, as HTLs [5–7]. However, these materials are expensive and degrade under air exposure [8,9]. Alternatively, metal oxides are shown to be promising candidates for HTL due to their low cost and improved stability [10]. Commonly used metal oxides are MoO_x and WO_3 [11,12], but these n-type semiconductors do not block electrons [13,14].

Among p-type HTLs, NiO$_x$ has been shown to have promising performance [15,16]. However, it suffers from low conductivity and high visible light absorption [17,18]. Therefore, developing new inorganic p-type HTLs is crucial to achieve highly efficient and stable devices.

Delafossite (AMO$_2$; A = Cu^{1+} or Ag^{1+} and M is a trivalent metal) compounds are p-type oxides and have drawn significant interest since Kawazoe et al. reported CuAlO$_2$ as a transparent oxide with room temperature conductivity up to 1 S cm^{-1} [19]. Since then, many Cu-based delafossites have been synthesized with M = Al, Sc, Cr, Mn, Fe, Co, Ga, and Rh [20]. CuCrO$_2$ (CCO) is particularly attractive due to its high conductivity [21]. Theoretical calculations and X-ray photoelectron spectroscopy (XPS) studies showed that Cu d states are dominant at the valence band maximum (VBM) and the intrinsic CCO conduction is through a Cu^I/Cu^{II} mixed valence hole mechanism [22,23]. The size of CCO nanoparticles can be very small, ~10 nm [24,25]. In solar energy harvesting, it was first used in p-type dye sensitized solar cells (DSSCs) [25]. Moreover, CCO has been shown as a promising HTL in OSCs and PSCs [24,26–29]. To further increase the hole concentration, efforts have been made to replace the trivalent Cr^{3+} cation with a divalent dopant such as Ni^{2+}, Mg^{2+}, or Zn^{2+} [30–32]. In particular, Mg has been shown to be an excellent dopant to increase CCO conductivity [22,31,33]. Theoretical calculations showed that Mg doping induces low-formation energy defects just above the VBM in CCO and introduces new Cu d states in the bandgap, thus leading to a Cu^I/Cu^{II} mixed valence and higher conductivity [22,33]. Compared with Be and Ca, defect states introduced by Mg are closest to the VBM, making it a more effective dopant [34]. Several experimental studies also confirm that Mg doping increases electrical conductivity [31,35–37]. In p-type DSSCs, Mg doped CCO (Mg:CCO) has performed superior to undoped CCO [38,39].

Based on our results of using undoped CCO as HTL in OSCs [24] and PSCs [27], we hypothesized that Mg:CCO could further improve the solar cell performance. Mg:CCO has not been applied as HTL in OSCs and, to our knowledge, there is only one report of using Mg:CCO as HTL in PSCs [40]. In this work, we examine the effect of Mg doping on CCO nanoparticles and their performance as HTLs in OSCs and PSCs. The CCO and Mg:CCO nanoparticles are synthesized by a hydrothermal method. The influence of Mg doping on structural, chemical, morphological, optical, and electronic properties of CCO films are carefully characterized by X-ray diffraction (XRD), energy-dispersive X-ray spectroscopy (EDX), transmission electron microscopy (TEM), dynamic light scattering (DLS), X-ray photoelectron spectroscopy (XPS), scanning electron microscopy (SEM), ultraviolet–visible (UV-vis) absorption spectrometry, photo-electron spectroscopy in air (PESA), and Kelvin probe (KP) techniques. Finally, spin-coated CCO and Mg:CCO nanoparticle films are used as HTLs in three different BHJ OSCs and methylammonium lead iodide (MAPbI$_3$) PSCs. Time-resolved photoluminescence (TRPL) is applied to probe charge transfer between MAPbI$_3$ and HTL, and XPS is used to examine elemental diffusion. This is the first work to apply Mg:CCO nanoparticle films as HTLs in OSCs and the first demonstration in a non-fullerene acceptor BHJ system.

2. Materials and Methods

The chemicals used in this study included copper(II) nitrate hemipentahydrate (Cu(NO$_3$)$_2$·2.5 H$_2$O, Alfa Aesar, ACS, 98.0–102.0%, Tewksbury, MA, USA), chromium(III) nitrate nonahydrate (Cr(NO$_3$)$_3$·9 H$_2$O, Alfa Aesar, 98.5%, Tewksbury, MA, USA), magnesium nitrate hydrate (Mg(NO$_3$)$_2$, Avocado Research Chemicals, 99.999%, London, United Kingdom), sodium hydroxide (NaOH, Sigma Aldrich, ≥97.0%, St. Louis, MO, USA), hydrochloric acid (HCl, Fisher Scientific, 37%, Houston, TX, USA), ethanol (EtOH, Fisher Chemical, anhydrous, Houston, TX, USA), 2-methoxyethanol (2-MOE, Acros Organics, 99+%, Houston, TX, USA), acetone (Fisher Chemical, certified ACS, Houston, TX, USA), 2-propanol (IPA, Fisher Chemical, certified ACS plus, Houston, TX, USA), poly(3-hexylthiophene-2,5-diyl) (P3HT, Rieke Metals, LLC, ≥96%, Lincoln, NE, USA), [6:6]-phenyl C61-butyric acid methyl ester (PC$_{61}$BM, Solenne BV, >99%, Groningen, The Netherlands), [6,6]-phenyl C71-butyric acid methyl ester (PC$_{71}$BM, Solenne BV, >99%, Groningen, The Netherlands), poly(5-bromo-4-(2-octyldodecyl)-selenophen2-yl)-5,6-difluorobenzothiadiazole-5,5′-bis-

(trimethylstannyl)-2,2′-bithiophene (PFBT2Se2Th), poly[4,8-bis(5-(2-ethylhexyl)thiophen-2-yl)benzo[1,2-b;4,5-b′]dithiophene-2,6-diyl-alt-(4-(2-ethylhexyl)-3-fluorothieno[3,4-b]thiophene-)-2-carboxylate-2-6-diyl)] (PTB7-Th, Luminescence Technology Corp., New Taipei City, Taiwan), 3,9-bis(2-methylene-(3-(1,1-dicyanomethylene)-indanone))-5,5,11,11-tetrakis(4-hexylphenyl)-dithieno[2,3-d:2′,3′-d′]-s-indaceno[1,2-b:5,6-b′]dithiophene (ITIC, 1-Material Inc., Dorval, QC, Canada), chlorobenzene (CB, Sigma-Aldrich, anhydrous, 99.8%, St. Louis, MO, USA), 1,4-dichlorobenzene (DCB, Sigma Aldrich, anhydrous, 99%, St. Louis, MO, USA), diphenyl ether (DPE, Acros Organics, 99%, Houston, TX, USA), chloroform (CF, Sigma-Aldrich, ACS reagent, ≥99.8%, St. Louis, MO, USA), lead(II) iodide (TCI, 99.99%, Portland, OR, USA), methylammonium iodide (Dyesol, Queanbeyan, NSW, Australia), potassium iodide (Alfa Aesar, 99.998%, Tewksbury, MA, USA), dimethylformamide and dimethyl sulfoxide (DMF and DMSO, Sigma Aldrich, anhydrous grade, St. Louis, MO, USA), C_{60} (Luminescence Technology Corp., >99.5%, New Taipei City, Taiwan), and bathocuproine (BCP) (Sigma Aldrich, sublimed grade, 99.99%, St. Louis, MO, USA). All chemicals were used without further purification. PFBT2Se2Th was prepared according to previous publication, via copolymerization of 4,7-Bis(5-bromo-4-(2-octyldodecyl)selenophen-2-yl)-5,6-difluorobenzothiadiazole (FBT2Se) and 5,5′-bis(trimethylstannyl)-2,2′-bithiophene (2Th) [41].

2.1. CuCrO2 (CCO) and Mg:CCO Preparation

2.1.1. Nanoparticle Synthesis

Mg:CCO nanoparticles with 0 at%, 5 at%, and 10 at% Mg doping levels were synthesized by a hydrothermal method as reported in previous literature [39]. First, 7.5 mmol $Cu(NO_3)_2 \cdot 2.5H_2O$ and stoichiometric amounts of $Cr(NO_3)_3 \cdot 9H_2O$ and $Mg(NO_3)_2$ were dissolved in 35 mL deionized (DI) water and stirred for 15 min at room temperature. Next, 2.5 g NaOH was added into the mixture and stirred for another 15 min at room temperature. The precursor solution was transferred into a 50 mL autoclave reactor (Col-Int Tech., Irmo, SC, USA), filled to 70% of its total volume. The hydrothermal reaction was carried out at 240 °C for CCO and 230 °C for Mg:CCO for 60 hours. Finally, the precipitate was washed using 2 M HCl and EtOH in sequence several times until the supernatant was colorless. After centrifuging, the mud was dried in a desiccator at room temperature overnight to obtain CCO or Mg:CCO powders.

2.1.2. Suspension Preparation

CCO and Mg:CCO nanoparticles were dispersed into 2-MOE to make 2 mg mL^{-1} suspensions for materials characterization and fabrication of P3HT:PC$_{61}$BM OSCs, PFBT2Se2Th:PC$_{71}$BM OSCs, PTB7-Th:ITIC OSCs, and MAPbI$_3$ PSCs. Just prior to film preparation, the suspensions were placed into a bath sonicator (Branson 3510, Plano, TX, USA) for 90 min, then filtered through a 1 µm PTFE filter (Thermos scientific, Titan3, Houston, TX, USA), re-sonicated for 90 min, and again filtered through a 0.45 µm PTFE filter (Biomed Scientific, Seattle, WA, USA).

2.1.3. Film Preparation

2 mg mL^{-1} CCO and Mg:CCO suspensions were drop cast on gold-coated silicon substrates for EDX (Zeiss Supra 40) and XPS and on Cu grids (Ted Pella, Inc., Redding, CA, USA) for TEM imaging. CCO and Mg:CCO suspensions were spin coated multiple times on glass substrates for UV-vis (Ocean Optics USB 4000, Largo, FL, USA), spectroscopic ellipsometry (J. A. Woollam M-2000DI, Lincoln, NE, USA), and PESA (RKI Instruments Model AC-2, Union City, CA, USA), and on ITO substrates for SEM (Zeiss Supra 40, Lewisville, TX, USA) and KP (KP Technology SKP 5050, Caithness, Scotland) measurements and solar cells. After spin coating, the CCO and Mg:CCO films were annealed on a hot plate (120–175 °C) for 5 min in air to evaporate the solvent.

2.2. Materials Characterizaton

The crystalline phases of the nanoparticles were characterized by XRD using a Rigaku Ultima III diffractometer (The Woodlands, TX, USA) with Cu Kα (λ = 1.5418 Å) radiation. Powder diffraction files (PDFs) were used to identify characteristic peaks in the XRD patterns. Polytype compositions, crystal size and lattice parameters of XRD patterns were performed using Profex (an open source XRD and Rietveld refinement software, Solothurn, Switzerland) and structure files for phase identification were downloaded from the Crystallography Open Database (COD). The film morphologies were examined using SEM. The experimental Mg doping concentration was quantified using EDX. Nanoparticle size was measured from TEM images from the Delong LVEM5 Benchtop Electron Microscope (Delong TEM, Montreal, QC, Canada) equipped with the Q-Capture Pro 7 software. Hydrodynamic sizes were measured by DLS using a Malvern Zetasizer Nano ZS instrument (Malvern, United Kingdom). The lattice fringe distances were determined from high resolution TEM (HR TEM) images of nanoparticles obtained using a JEOL JEM2100 TEM (Peabody, MA, USA). The elemental compositions and chemical states of the films were analyzed by XPS, using a PHI 5000 Versa Probe II equipped with an Al Kα source and a hemispherical analyzer. XPS data were taken at a 45° takeoff angle with a pass energy of 23.5 eV. Optical transmission of the films was characterized by UV-vis over the wavelength range from 178 to 890 nm. The band gap energy was determined from Tauc plots of the UV-vis absorbance data. Film thickness was obtained by ellipsometry at 55°, 65°, and 75° incident angles over the wavelength range from 280 to 1690 nm. Ionization energy was measured from 4.7 to 5.8 eV with a 0.05 eV energy step using PESA with deuterium lamp intensity at 100 nW. The work function was measured using a KP apparatus (SKP5050, KP Technology) referenced to Au at 5.15 eV.

2.3. Solar Cell Fabrication and Testing

2.3.1. P3HT:PC$_{61}$BM OSCs

P3HT:PC$_{61}$BM devices were fabricated on patterned ITO substrates (Xinyan Technology Ltd., Kwun Tong, Hong Kong, 15 Ω sq^{-1}). The substrates were rinsed using soapy water, acetone, and IPA, followed by UV-ozone treatment for 20 min. 2 mg mL^{-1} CCO and Mg:CCO suspensions were spin coated at 2000 rpm for 30 s on top of the ITO substrates. The thickness of CCO and Mg:CCO nanoparticle films are ~18 nm; see Section 3.2 for details. 23 mg mL^{-1} P3HT and 23 mg mL^{-1} PC$_{61}$BM were dissolved in CB and stirred at 50 °C overnight. The P3HT:PC$_{61}$BM active layer (~200 nm thick) was made by dispensing 35 µL P3HT:PC$_{61}$BM solution on a spinning substrate at 1200 rpm for 60 s, followed by annealing at 170 °C in N$_2$ for 10 min. Finally, 7 nm Ca and 100 nm Al were sequentially evaporated on top of the active layer. The current-voltage (J-V) measurements were carried out using a 2635A Keithley low-noise sourcemeter under AM 1.5G 100 mW cm^{-2} illumination from a class AAA solar simulator (Abet Technologies) in a nitrogen filled glovebox. The diode area is 0.11 cm^2 and the aperture area is 0.049 cm^2.

2.3.2. PFBT2Se2Th:PC$_{71}$BM OSCs

PFBT2Se2Th:PC$_{71}$BM devices were fabricated and tested similarly to the description in Section 2.3.1 unless otherwise noted. Six mg mL^{-1} PFBT2Se2Th and 12 mg mL^{-1} PC$_{71}$BM were dissolved in DCB with 5 vol % DPE and stirred at 100 °C overnight. This solution and ITO/CCO or Mg:CCO substrates were preheated at 100 °C. The PFBT2Se2Th:PC$_{71}$BM active layer (~120 nm thick) was made by first dispensing 50 µL of PFBT2Se2Th:PC$_{71}$BM solution on the substrate and then immediately starting spinning at 1200 rpm for 60 s, followed by drying in a vacuum chamber for 2 min.

2.3.3. PTB7-Th:ITIC OSCs

PTB7-Th:ITIC devices were fabricated and tested similarly to the description in Section 2.3.1 unless otherwise noted. PTB7-Th and ITIC were blended in a 1:1 weight ratio, dissolved in a mixed solution (CB with 3 vol% CF) at a total concentration of 20 mg mL^{-1} and stirred at room temperature overnight.

The PTB7-Th:ITIC active layer (~80 nm thick) was made by first dispensing 40 µL PTB7-Th:ITIC solution on the substrate and then immediately starting spinning at 1250 rpm for 60 seconds.

2.3.4. MAPbI$_3$ PSCs

MAPbI$_3$ PSCs were fabricated by spin coating the MAPbI$_3$ layer (~450 nm thick) according to an antisolvent-washing recipe first described by Ahn et al. [42], and thereafter thermally evaporating a C$_{60}$/BCP electron transport layer and then Ag electrodes. The details of the deposition of each of these layers are exactly as described in our previous report [27], except that the antisolvent wash during MAPbI$_3$ deposition was carried out 11–12 s after starting the spin recipe. After fabrication, the devices were encapsulated with Ossila E131 UV-cure epoxy and a glass coverslip. J.–V curves were measured using a Keithley 2401 sourcemeter and an Oriel solar simulator. The lamp intensity was initially calibrated to 1 sun using a reference Si solar cell from Newport Corp., and maintained at that intensity during the measurements by a reference photodiode. The aperture area of the PSCs is 0.1 cm^2, as defined by a shadow mask. (J,V) points were collected by sampling the current 1 s after the bias was applied. AC external quantum efficiency measurements were performed using an Enlitech QE-R instrument equipped with a monochromated Xe lamp optically chopped at 165 Hz, and without applied electrical bias. XPS, using a Kratos Analytical Axis Ultra spectrometer equipped with a monochromated Al Kα source, was performed on ITO/HTL/MAPbI$_3$ films to determine whether diffusion of Cu, Cr, or Mg from the HTL resulted in detectable levels of these elements at the surface of the perovskite film. TRPL experiments comparing ITO/MAPbI$_3$ and ITO/HTLs/MAPbI$_3$ structures were performed using a microscope-based time-resolved system [43]. Samples were excited by 405 nm/120 fs optical pulses at 7.6 MHz repetition rate produced by doubling the fundamental frequency of the Mira 900 laser and followed by pulse-picking (1 out of 10 pulses) via the acousto-optical modulator (NEOS Technologies). Excitation of 1 µW was focused on the sample via 0.6 NA objective, which also ensured a high photon collection efficiency to obtain PL signals. The collected emission was passed through a spectrometer and directed either to a CCD camera for PL spectral analysis or to a sensitive photon detector (MicroPhoton Devices MPD 50) for the wavelength-dependent PL lifetime measurements. PL decay curves were collected via the time-correlated single photon counting performed on board of Pico300E photon counting hardware (PicoQuant GmbH). The overall time resolution was better than 200 ps.

3. Results and Discussion

3.1. Structural, Compositional, and Morphological Characterizations

Figure 1a shows XRD patterns of CCO (black curve), 5% Mg:CCO (red curve), and 10% Mg:CCO (blue curve) powders, respectively. XRD peaks are indexed as a mixture of two CCO polytypes, rhombohedral R3m (3R-CCO, pink sticks), and hexagonal P63/mmc (2H-CCO, purple sticks), for all three compounds. No impurity phases are detected. For the (110) reflection at ~62.0°, Mg doping results in a ~0.05° peak shift to the lower angle, indicating a larger lattice spacing. The (004) reflection at 31.4° exhibits broadening with Mg doping, indicating decreased crystal size along the c axis (Figure 1a and Table 1). Rietveld refinement was carried out in order to quantitatively determine the polytype composition, crystal size, and lattice parameters for each XRD pattern. Figure 1b shows the experimental (blue solid circle), calculated (red curve), and difference between experimental and calculated (grey curve) patterns for CCO, 5% Mg:CCO, and 10% Mg:CCO. Table 1 shows the results extracted from the Rietveld refinement. The polytype compositions for all three compounds are found to be ~60 ± 3% for 3R and ~40 ± 3% for 2H. The crystal sizes decrease monotonically from CCO to 10% Mg:CCO, from 7.8 to 4.5 nm for the 2H polytype calculated from the (004) reflection and from 9.6 to 8.7 nm for the 3R polytype calculated from the (110) reflection. However, the crystal size increases monotonically from 10.2 nm for CCO to 13.1 nm for 10% Mg:CCO in the (110) reflection for the 2H polytype. The tradeoff of the size changes for both polytypes leads to similar widths of the

(110) reflection independent of Mg doping. Since delafossite nanocrystals often exhibit anisotropic morphology [25,27], it is reasonable that size changes differ for the (004) and (110) reflections. Similar orientation dependent size variation in Mg:CCO was reported by Bywalez et al. [35]. They attributed the decrease of crystal sizes along the c axis to Mg^{2+} obstructing growth of the delafossite crystal structure and stabilizing the spinel phase. However, there is no indication that this phase exists in our samples.

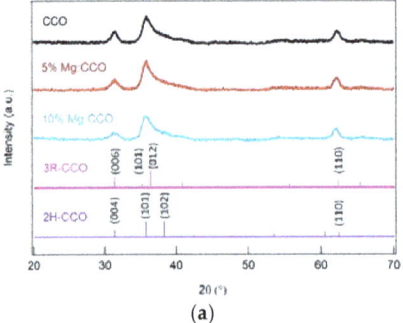 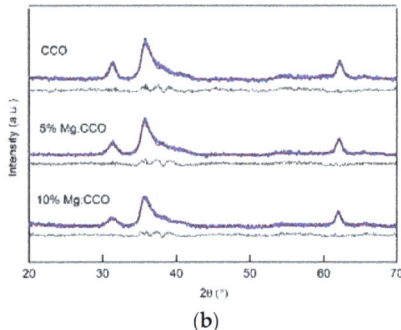

(a) (b)

Figure 1. (a) X-ray diffraction (XRD) patterns of $CuCrO_2$ (CCO) (black curve), 5% Mg:CCO (red curve), and 10% Mg:CCO (blue curve) powders. Two CCO polytypes, 3R-CCO (Powder diffraction files (PDF) #39-0247, pink sticks) and 2H-CCO (PDF#89-6743, purple sticks) are detected in all three XRD patterns. Prominent reflections are indexed between 30° to 40°, and ~62.0°. (b) Rietveld refinement of XRD patterns. The experimental (blue solid circle), calculated (red curve) and, difference (grey curve) patterns are shown. Structural files for the refinement are 3R-CCO (Crystallography Open Database (COD) No. 8104066) and 2H-CCO (COD No. 8104065).

Table 1. Rietveld refinement results for CCO, 5% Mg:CCO, and 10% Mg:CCO.

Sample	CCO	5% Mg:CCO	10% Mg:CCO
Polytype composition (%)	3R-CCO = 59.1 ± 3.0	3R-CCO = 59.5 ± 2.6	3R-CCO = 56.0 ± 2.7
	2H-CCO = 40.9 ± 3.0	2H-CCO = 40.5 ± 2.6	2H-CCO = 44.0 ± 2.7
R_{wp} (%) [1]	13.3	13.7	13.4
R_{exp} (%) [2]	9.0	9.0	9.1
R_p (%) [3]	9.0	10.9	10.5
χ^2	2.2	2.3	2.2
Crystal size (nm) based on (004)	7.8 ± 0.4 (2H-CCO)	5.6 ± 0.3 (2H-CCO)	4.5 ± 0.3 (2H-CCO)
Crystal size (nm) based on (110)	9.6 ± 0.9 (3R-CCO)	9.4 ± 0.9 (3R-CCO)	8.7 ± 1.0 (3R-CCO)
	10.2 ± 1.0 (2H-CCO)	12.3 ± 1.3 (2H-CCO)	13.1 ± 1.5 (2H-CCO)
Lattice parameter a [4] and c [5] (Å) for 3R-CCO	a = 2.99	a = 2.99	a = 3.00
	c = 17.44	c = 17.44	c = 17.44
Lattice parameter a and c (Å) for 2H-CCO	a = 2.99	a = 2.99	a = 3.00
	c = 11.43	c = 11.44	c = 11.46

[1] R_{wp} is the weighted profile R-factor and the squared R_{wp} is equal to the weighted sum of squared difference between the experimental and calculated intensity values over the weighted sum of squared experimental intensity values [44]. [2] R_{exp} is the expected R-factor and the squared R_{exp} is equal to the number of data points over the weighted sum of squared experimental intensity values [44]. [3] R_p is the profile R-factor and is equal to the weighted sum of difference between the experimental and calculated intensity values over the weighted sum of experimental intensity values [45]. [4 and 5] a and c are the in-plane and out-of-lane lattice constants in the unit cell.

The lattice parameters for 3R-CCO and 2H–CCO polytypes were similar for CCO and 5% Mg:CCO. However, for 10% Mg:CCO, a larger a lattice parameter in both phases (3.00 Å versus 2.99 Å) and an increase in the c lattice parameter in the 2H–CCO phase (11.46 Å versus 11.43 Å) were observed. This lattice expansion was consistent with Mg substituting on the Cr site, rather than the Cu site [36], because the ionic radius of Mg^{2+} (0.72 Å) is larger than that of Cr^{3+} (0.62 Å) and smaller than that of

Cu$^+$ (0.77 Å). This result is consistent with the bond length increase between Cr and O sites after Mg doping predicted from theoretical calculations [33].

In order to measure Mg concentration in CCO, EDX was performed on 5% Mg:CCO and 10% Mg:CCO. The insets of Figure 2a,b show that Mg is present and distributed uniformly in the Mg:CCO films. Mg/(Mg+Cr) represents the Mg concentration in the Mg:CCO films, which is calculated by atomic number effects (Z), absorption (A), and fluorescence (F) method from EDX spectra in Figure 2 [46]. Table 2 shows that the averaged Mg concentration in 5% Mg:CCO and 10% Mg:CCO measured from EDX is 4.0% and 9.8%, respectively. A possible Mg doping process is proposed similarly to the CCO formation mechanism as described by Miclau et al. [47]. During the hydrothermal synthesis of Mg:CCO nanoparticles, Cu^{1+}, Cr^{3+}, and Mg^{2+} ions can form Cu(OH)$_2^-$, Cr(OH)$_4^-$, and Mg(OH)$_2$, respectively, at alkaline pH environment according to equations (1-3) below. Mg:CCO nanoparticles can then be formed from the metal hydroxides according to Equation (4) [48–50]. The formation process of Mg:CCO nanoparticles are given in the following equations:

$$Cu^{1+} + 2H_2O \rightarrow Cu(OH)_2^- + 2H^+ \tag{1}$$

$$Cr^{3+} + 4H_2O \rightarrow Cr(OH)_4^- + 4H^+ \tag{2}$$

$$Mg^{2+} + 2OH^- \rightarrow Mg(OH)_2 \tag{3}$$

$$Cu(OH)_2^- + (1-x)\cdot Cr(OH)_4^- + x\cdot Mg(OH)_2 + 2H^+ \rightarrow CuCr_{1-x}Mg_xO_2 + 4H_2O - 2x\cdot (OH)^- \tag{4}$$

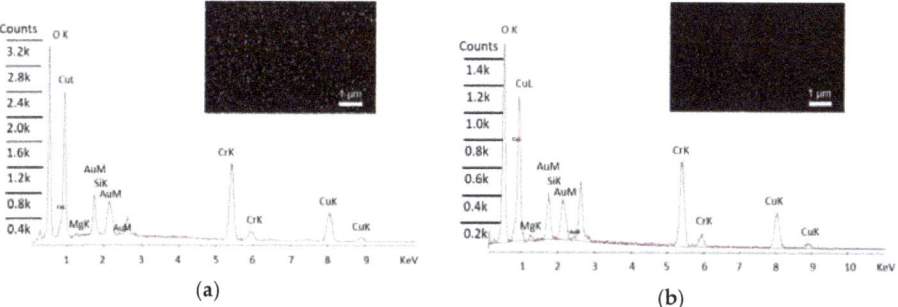

Figure 2. Energy-dispersive X-ray spectroscopy (EDX) spectra for (a) 5% Mg:CCO and (b) 10% Mg:CCO. Mappings of Mg element (inset) show uniform distribution.

Table 2. Measured Mg doping concentration of CCO, 5% Mg:CCO, and 10% Mg:CCO films and average transmission electron microscopy (TEM) nanoparticle sizes of CCO, 5% Mg:CCO, and 10% Mg:CCO nanoparticles.

Sample	CCO	5% Mg:CCO	10% Mg:CCO
Mg/(Mg+Cr) (%) [1]	0	4.0 ± 0.2	9.8 ± 1.3
Nanoparticle size (nm) [2]	10.3 ± 2.1	8.2 ± 2.1	9.8 ± 3.0

[1] Mg concentration is averaged over five EDX measurements. [2] Nanoparticle size is calculated from TEM images and mean size for each sample is averaged over 50 individual nanoparticles (Figure 3).

TEM images of CCO (Figure 3a), 5% Mg:CCO (Figure 3b), and 10% Mg:CCO (Figure 3c) show the nanoparticles exist in individual or small clusters as well as large agglomerates. We only use individual or double nanoparticles (indicated by white circles) to determine particle sizes. The average nanoparticle size for CCO, 5% Mg:CCO, and 10% Mg:CCO is 10.3 ± 2.1, 8.2 ± 2.1, and 9.8 ± 3.0 nm, respectively. The size trend according to TEM results differs slightly from that of Rietveld-refined XRD data. One difference is that the particle size determined from XRD is analyzed for specific reflection and polytype (Table 1), while TEM images are two-dimensional projections of nanoparticles with random orientation. To examine the TEM size results in details, Figure 4a shows box plots of TEM particle sizes for CCO (black), 5% Mg:CCO (red), and 10% Mg:CCO (blue) nanoparticles. The ranges of CCO and 5% Mg:CCO nanoparticle sizes are smaller than that of 10% Mg:CCO nanoparticles. It is clear that 90% of the CCO nanoparticles are larger than 8 nm. In contrast, significant fractions of both types of Mg:CCO nanoparticles are between 6 and 8 nm. Thus, the statistics of TEM results show that Mg:CCO samples have greater numbers of smaller-sized particles, although 10% Mg doping appears to broaden the distribution. Considering the particle size results from XRD and TEM, overall Mg doping decreases CCO nanoparticle sizes because the XRD results show the size decreases along the c axis and TEM results show greater numbers of smaller sized Mg:CCO particles.

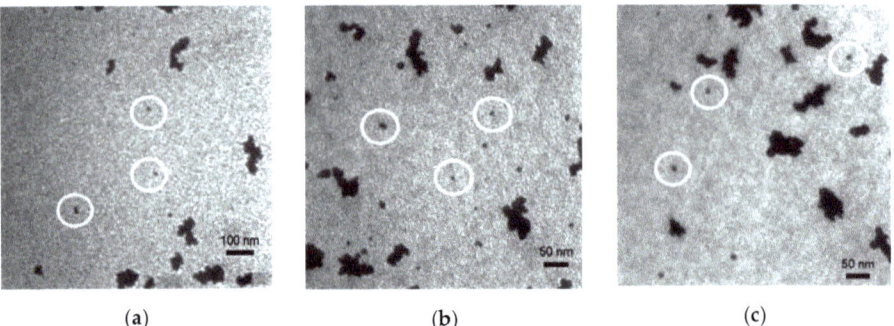

Figure 3. TEM images of (**a**) CCO, (**b**) 5% Mg:CCO, and (**c**) 10% Mg:CCO nanoparticles. Individual and double nanoparticles, as shown inside white circles in (**a**)–(**c**) are used to calculate size distributions shown in Table 2 and Figure 4a.

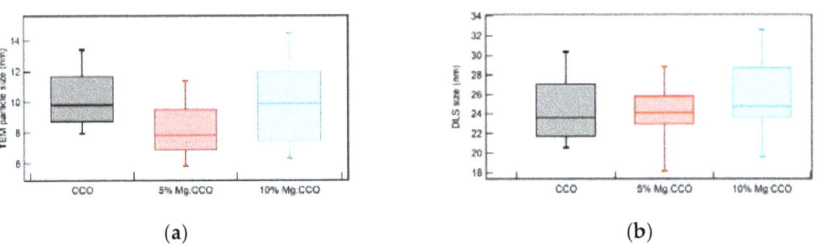

Figure 4. Box plots of (**a**) TEM-determined particle sizes from Figure 3 and (**b**) dynamic light scattering (DLS)-determined sizes for CCO (black color), 5% Mg:CCO (red color), and 10% Mg:CCO (blue color) nanoparticles. The percentiles are set to 90% whisker top, 75% box top, 25% box bottom, and 10% whisker bottom for each data set. In (**a**), ~50 individual nanoparticles are used in each data set. In (**b**), ~12 batches of DLS measurements are used in each data set.

Figure 4b shows box plots of DLS sizes for CCO (black), 5% Mg:CCO (red), and 10% Mg:CCO (blue) nanoparticles dispersed in 2-MOE. For all nanoparticles, the 25–75% distributions are between 20 to 30 nm, indicating nanoparticles disperse well in 2-MOE. There are no significant differences among the three doping concentrations due to possibly similar hydrodynamic layer thickness. This is expected because the hydrodynamic layer is determined by solution ionic strength and hydrodynamic size and is often larger than dry particle size [51,52]. Since the Mg concentration in these nanoparticles is low, it is not surprising that hydrodynamic sizes for all samples are similar.

The HR TEM image of 5% Mg:CCO nanoparticles shows clear lattice fringes (Figure 5). The lattice spacing of 2.47 Å corresponds to the (012) reflection for 3R–CCO polytype. The lattice spacing of 2.33 Å corresponds to the (102) reflection for 2H–CCO polytype. No other lattice spacings corresponding to impurity phases are detected, consistent with XRD results.

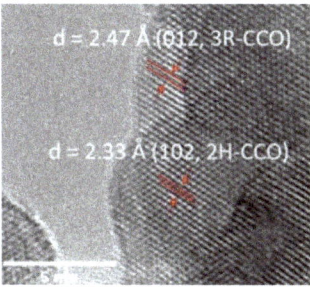

Figure 5. High resolution TEM (HR TEM) image of 5% Mg:CCO nanoparticle. Lattice spacings corresponding to the (012) reflection in 3R-CCO and the (102) reflection in 2H-CCO polytypes are indicated.

XPS studies were carried out in order to confirm the oxidation states of Cu, Cr, and Mg in the Mg:CCO powders. XPS data was analyzed using PHI Multipak software and peak fitting was done using a Gaussian–Lorentzian profile after a Shirley type background subtraction [53]. The binding energy was shifted using the valence band edge. The measured (cross symbol) and fitted (solid curve) XPS spectra of Cu $2p_{3/2}$, Cr $2p_{3/2}$, Mg 1s, and O 1s core levels for 5% Mg:CCO (red color) and 10% Mg:CCO (blue color) nanoparticles are shown in Figure 6. Deconvolution of the Cu $2p_{3/2}$ spectrum for both 5% and 10% Mg:CCO (Figure 6a) results in two peaks at 934.6 eV and 932.3 eV corresponding to binding energies of $Cu(OH)_2$ and Cu^{1+}, respectively, consistent with the literature [54,55]. The Cr $2p_{3/2}$ spectrum (Figure 6b) can be fitted to two peaks at 577.3 eV and 576.5 eV corresponding to binding energies of Cr^{3+} as hydroxide and Cr^{3+} as oxide, respectively [37,56]. These are similar to the binding energy peak positions of Cu^{1+} and Cr^{3+} oxide of undoped CCO nanoparticles reported in the literature [24]. Figure 6c shows the Mg 1s spectra, wherein the peak is located at binding energy of 1303.1 eV, corresponding to the Mg^{2+} oxidation state [39]. The O 1s spectrum (Figure 6d) shows peaks corresponding to lattice oxygen (O_I) at 529.9 eV and hydroxyl groups (O_{II}) at 531.5 eV for both 5% Mg:CCO and 10% Mg:CCO nanoparticles. A small-intensity peak at 533 eV corresponding to adsorbed water for 5% Mg:CCO nanoparticles is observed [54,57].

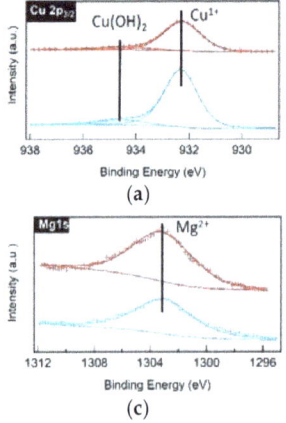
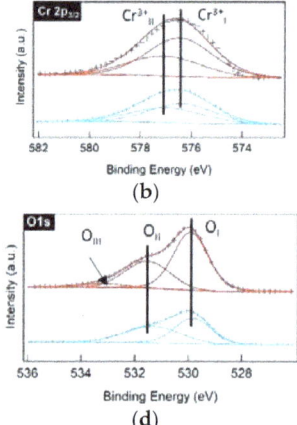

Figure 6. X-ray photoelectron spectroscopy (XPS) spectra of (**a**) Cu 2p$_{3/2}$, (**b**) Cr 2p$_{3/2}$, (**c**) Mg 1s, and (**d**) O 1s orbitals for 5% Mg:CCO (red) and 10% Mg:CCO (blue) nanoparticles. The measured XPS spectra are represented by cross symbols. The fitted XPS spectra are represented by solid curves. The black lines show the binding energies for Cu(OH)$_2$, Cu^{1+}, Cr$^{3+}{}_I$, Cr$^{3+}{}_{II}$, Mg^{2+}, O$_I$, O$_{II}$, and O$_{III}$. Cr$^{3+}{}_I$ represents Cr^{3+} as oxide, Cr$^{3+}{}_{II}$ is Cr^{3+} as hydroxide; O$_I$ represents the lattice oxygen, O$_{II}$ is hydroxyl species, and O$_{III}$ is adsorbed water.

3.2. Optical and Electronic Characterizations

The thickness of CCO and Mg:CCO nanoparticle films are controlled by the number of coating cycles that were performed during deposition [27]. Figure 7a–c shows SEM images for CCO, 5% Mg:CCO, and 10% Mg:CCO films; no regions of bare substrate are seen for all films. Figure 8a shows the UV-vis absorbance and transmission (inset) spectra of well-covered CCO (black), 5% Mg:CCO (red), and 10% Mg:CCO (blue) films. The absorbance values at 300 nm lie between 0.21 and 0.22 for all films, which are highly transparent (transmission > 90%) in the visible region. All three films are 18 nm thick as determined by ellipsometry. The direct band gap (E_g) is extrapolated from the Tauc plot (Figure 8b). The average values of the direct E_g are 3.27 ± 0.02 eV, 3.25 ± 0.03 eV, and 3.27 ± 0.03 eV for CCO, 5% Mg:CCO, and 10% Mg:CCO, respectively (Table 3). These values are the same within the uncertainty of the measurement (~0.03 eV). Thus, Mg doping does not affect the direct E_g of CCO. The E_g for pure CCO is consistent with our previous result [27].

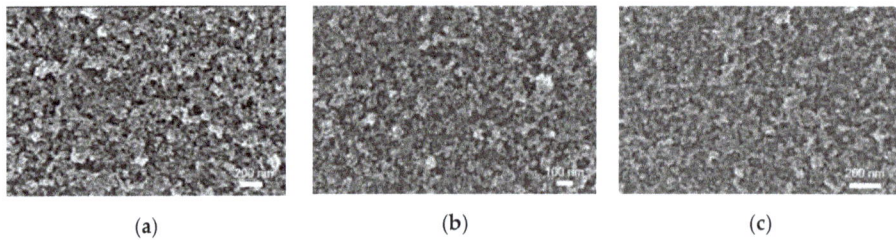

Figure 7. SEM images of (**a**) CCO, (**b**) 5% Mg:CCO, and (**c**) 10% Mg:CCO films on ITO substrates.

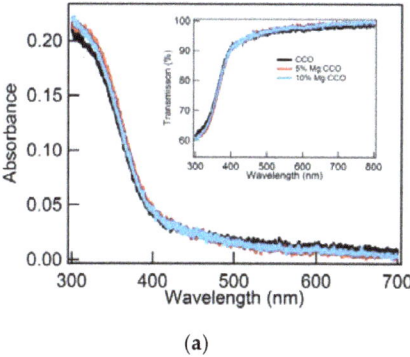
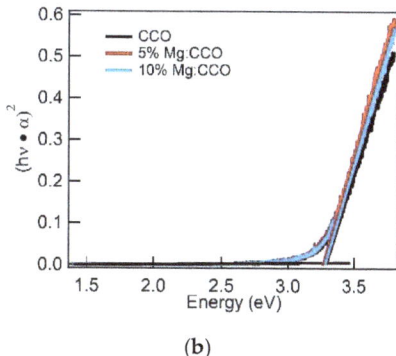

Figure 8. (a) Ultraviolet–visible (UV-vis) absorbance and transmission (inset) spectra and (b) Tauc plots and linear fits of the band edge (straight lines) for CCO (black curve), 5% Mg:CCO (red curve), and 10% Mg:CCO (blue curve) films.

Table 3. Thickness and direct E_g of CCO, 5% Mg:CCO, and 10% Mg:CCO films with ~0.22 absorbance at 300 nm wavelength.

Sample	Thickness (nm)	Direct E_g (eV) [1]	$WF_{median} - IE_{median}$ (eV)
CCO	18	3.27 ± 0.02	0.08
5% Mg:CCO	18	3.25 ± 0.03	0.09
10% Mg:CCO	18	3.27 ± 0.03	0.16

[1] Direct E_g is averaged over three measurements.

Figure 9a shows box plots of the work function (WF) for CCO (black), 5% Mg:CCO (red), and 10% Mg:CCO (blue) films. The median WF values of CCO, 5% Mg:CCO, and 10% Mg:CCO films are 5.19, 5.17, and 5.22 eV, respectively. The 50th percentile WF value for the 10% Mg:CCO is higher than that of CCO and 5% Mg:CCO. The 5% Mg:CCO films exhibit the largest spread with a long tail to the large WF than CCO films. Thus, Mg:CCO films generally appear to have higher WF values than CCO films, although the difference is below the level of statistical significance. Figure 9b shows box plots of the ionization energy (IE) for CCO (black), 5% Mg:CCO (red), and 10% Mg:CCO (blue) films. The median IE values of CCO, 5% Mg:CCO, and 10% Mg:CCO films are 5.11, 5.08, and 5.06 eV, respectively. The energy step for the IE measurement is 0.05 eV. The 50-percentile IE value decreases monotonically with increasing Mg concentration. Furthermore, the measured IE value is consistent with the IE of 5.1 eV from previous band structure calculations [27]. As shown in Figure 9, the overall WF values are larger than IE values, especially for Mg:CCO films. Thus, these films are p-type degenerately doped. The difference between WF and IE values (WF – IE) increases with Mg concentration from 0.08 eV for CCO to 0.16 eV for 10% Mg:CCO (Table 3), indicating that Mg:CCO films may have higher conductivity, consistent with previous results [35,36].

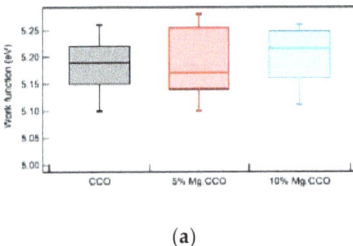
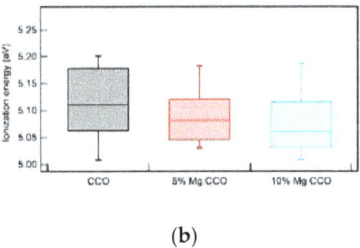

(a) (b)

Figure 9. Box plots of (**a**) work function and (**b**) ionization energy for CCO (black color), 5% Mg:CCO (red color), and 10% Mg:CCO (blue color) films. The percentiles are set to 90% whisker top, 75% box top, 25% box bottom, and 10% whisker bottom for each data set. In (**a**), 20 batches of WF measurements are used in each data set. In (**b**), 13 batches of IE measurements are used in each data set.

3.3. CCO and Mg:CCO as HTLs in OSCs and PSCs

Figure 10a and Table 4 show the results of P3HT:PC$_{61}$BM devices with CCO (black), 5% Mg:CCO (red), and 10% Mg:CCO (blue) HTLs. The average J_{sc} of P3HT:PC$_{61}$BM devices is higher with Mg:CCO HTL (from 6.94 mA cm^{-2} for CCO to ~7.05 mA cm^{-2} for Mg:CCO). The average V_{oc} of Mg:CCO is also higher than that of undoped CCO HTL (~0.582 V versus 0.570 V). However, both Mg:CCO devices exhibit lower average FF (0.642 for 5% doping and 0.666 for 10% doping versus 0.685 for no doping). The tradeoff of the three parameters leads to similar PCE values for all devices independent of Mg doping. We note that the variation among different diodes is larger in J_{sc} than V_{oc} or FF, which is typical of OPV devices. Nonetheless, there is a systematic trend of average J_{sc} increase with Mg doping.

Figure 10b and Table 4 show the results of PFBT2Se2Th:PC$_{71}$BM devices with CCO (black), 5% Mg:CCO (red), and 10% Mg:CCO (blue) HTLs. The average V_{oc} and FF are similar in all devices with values of ~0.665 V and ~0.685, respectively. The average J_{sc} of the devices increases monotonically from 10.50 mA cm^{-2} for undoped CCO HTL to 10.88 mA cm^{-2} for 10% Mg:CCO HTL. Thus, the PCE values of PFBT2Se2Th:PC$_{71}$BM devices are higher when Mg:CCO, instead of undoped CCO, is used as the HTL. Among different diodes, the variation of J_{sc} is larger than that of V_{oc} or FF. Moreover, Mg:CCO devices have even larger variation of J_{sc}. However, a similar systematic trend of increasing average J_{sc} as P3HT:PC$_{61}$BM devices is observed in PFBT2Se2Th:PC$_{71}$BM devices.

Figure 10c and Table 4 show the results of PTB7-Th:ITIC devices with CCO (black), 5% Mg:CCO (red), and 10% Mg:CCO (blue) HTLs. The average J_{sc} of the devices increases monotonically from 11.55 mA cm^{-2} for undoped CCO HTL to 12.02 mA cm^{-2} for 10% Mg:CCO HTL. The average V_{oc} and FF are highest for the 5% Mg:CCO in this batch of devices, but they do not depend on Mg doping in other batches. Generally, the PCE values of PTB7-Th:ITIC devices are higher when using Mg:CCO as the HTL due to the increase in J_{sc}. This work is the first using CCO and Mg:CCO as HTL for BHJ OSCs with a non-fullerene acceptor.

Figure 10d and Table 4 show the results of MAPbI$_3$ PSCs with CCO (black), 5% Mg:CCO (red), and 10% Mg:CCO (blue) HTLs under forward (solid lines) and reverse scans (dashed lines). Only slight hysteresis is seen, indicating minimal trap states at the CCO or Mg:CCO/perovskite interface. Under forward scan, the average V_{oc} increases monotonically from 0.985 V for the undoped CCO HTL to 1.007 V for the 10% Mg:CCO HTL. Similar trends are observed in the J_{sc} (from 18.91 mA cm^{-2} for the undoped CCO HTL to 19.40 mA cm^{-2} for the 10% Mg:CCO HTL) and FF (from 0.678 for the undoped CCO HTL to 0.703 for the 10% Mg:CCO HTL). Overall, the PCE of the devices improves monotonically from 12.64% for the undoped CCO HTL to 13.73% for the 10% Mg:CCO HTL. We note that the variation among different diodes is large for all the parameters. However, there are systematic trends of increases among the average J_{sc}, V_{oc}, and FF with Mg doping. Under reverse scan, the average J_{sc} increases monotonically from 18.70 mA cm^{-2} for the undoped CCO HTL to 19.37 mA cm^{-2} for the 10% Mg:CCO HTL. The FF of the devices using the 5% Mg:CCO and 10% Mg:CCO HTLs were similar,

0.719, but for the undoped CCO HTL, a lower *FF* (0.697 versus 0.719) was observed. The V_{oc} is similar in all devices with values of ~1.01 V. Overall, the *PCE* of the devices improves monotonically from 13.19% for the undoped CCO HTL to 14.12% for the 10% Mg:CCO HTL. As in the forward scan data, despite the variation among different diodes, there is a systematic trend of increasing average J_{sc} with Mg doping. Jeong et al. observed that Mg:CCO produces PSCs with a slightly higher J_{sc} and V_{oc}, but a lower *FF*, resulting in no improvement in the *PCE*; however, they did not report Mg concentration [40]. The inset in Figure 8d shows the external quantum efficiency (EQE) at wavelength ranging from 300 to 800 nm for CCO (black), 5% Mg:CCO (red), and 10% Mg:CCO (blue) HTLs. A broadband increase in EQE is observed with Mg doping, consistent with the increases of average J_{sc} in forward and reverse *J-V* scans.

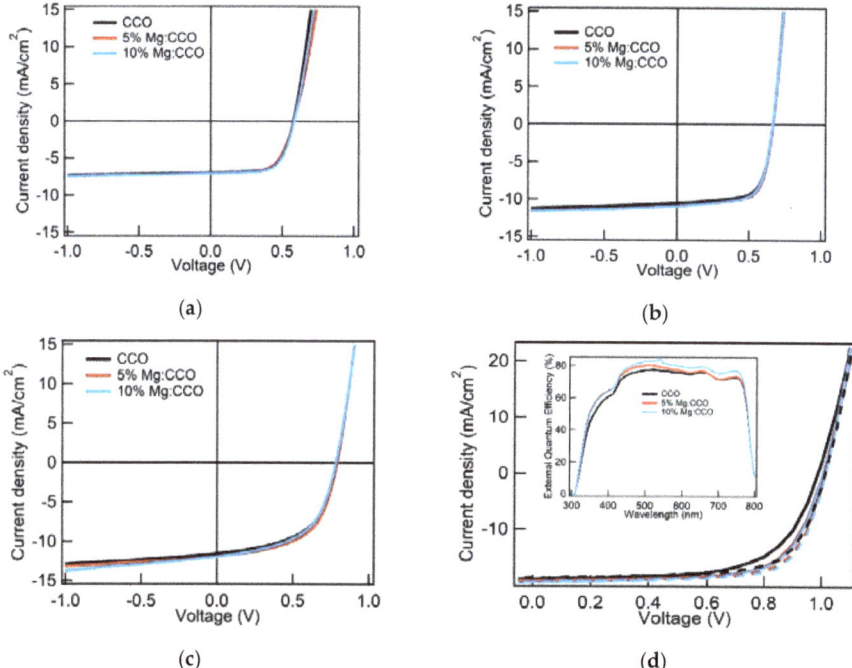

Figure 10. Average *J-V* curves (number of devices for each system is given in the footer of Table 4) of (**a**) P3HT:PC$_{61}$BM OSCs, (**b**) PFBT2Se2Th:PC$_{71}$BM OSCs, (**c**) PTB7-Th:ITIC OSCs, and (**d**) MAPbI$_3$ PSCs measured in AM 1.5G 100 mW cm^{-2} illumination with CCO (black curve), 5% Mg:CCO (red curve), and 10% Mg:CCO (blue curve) hole transport layers (HTLs). In (d), solid *J-V* curves are measured under forward scan, dashed *J-V* curves are measured under reverse scan, and the inset is the external quantum efficiency (EQE) measurements of representative MAPbI$_3$ cells with CCO (black curve), 5% Mg:CCO (red curve), and 10% Mg:CCO (blue curve) HTLs.

Table 4. The device parameters of OSCs and MAPbI$_3$ PSCs with CCO, 5% Mg:CCO, and 10% Mg:CCO HTLs.

Device Type [1]	HTL_Type	J_{sc} (mA cm^{-2})	V_{oc} (V)	FF	PCE (%)
P3HT:PC$_{61}$BM	CCO	6.94± 0.15	0.570 ± 0.000	0.685 ± 0.008	2.71 ± 0.06
	5% Mg:CCO	7.04 ± 0.11	0.583 ± 0.005	0.642 ± 0.022	2.63 ± 0.08
	10% Mg:CCO	7.06 ± 0.11	0.581 ± 0.007	0.666 ± 0.017	2.73 ± 0.03
PFBT2Se2Th:PC$_{71}$BM	CCO	10.50 ± 0.29	0.666 ± 0.007	0.684 ± 0.014	4.78 ± 0.18
	5% Mg:CCO	10.77 ± 0.61	0.664 ± 0.007	0.689 ± 0.011	4.93 ± 0.26
	10% Mg:CCO	10.88 ± 0.50	0.665 ± 0.007	0.678 ± 0.011	4.91 ± 0.27
PTB7-Th:ITIC	CCO	11.55 ± 0.17	0.786 ± 0.007	0.548 ± 0.010	4.97 ± 0.14
	5% Mg:CCO	11.87 ± 0.15	0.793 ± 0.005	0.559 ± 0.003	5.26 ± 0.08
	10% Mg:CCO	12.02 ± 0.27	0.785 ± 0.007	0.541 ± 0.011	5.11 ± 0.22
MAPbI$_3$ PSC (forward scan)	CCO	18.91± 0.43	0.985 ± 0.058	0.678 ± 0.025	12.64 ± 0.99
	5% Mg:CCO	19.26 ± 0.54	1.003 ± 0.010	0.696 ± 0.023	13.45 ± 0.44
	10% Mg:CCO	19.40 ± 0.39	1.007 ± 0.014	0.703 ± 0.018	13.73 ± 0.34
MAPbI$_3$ PSC (reverse scan)	CCO	18.70± 0.31	1.012 ± 0.006	0.697 ± 0.031	13.19 ± 0.71
	5% Mg:CCO	19.20 ± 0.41	1.011 ± 0.004	0.719 ± 0.012	13.96 ± 0.33
	10% Mg:CCO	19.37 ± 0.35	1.014 ± 0.006	0.719 ± 0.012	14.12 ± 0.28

[1] For P3HT:PC$_{61}$BM OSCs, 12, nine, and eight devices were measured for CCO, 5% Mg:CCO, and 10% Mg:CCO HTL, respectively. For PFBT2Se2Th:PC$_{71}$BM OSCs, eight, seven, and 11 devices were measured for CCO, 5% Mg:CCO, and 10% Mg:CCO HTL, respectively. For PTB7-Th:ITIC OSCs, 10, eight, and nine devices were measured for CCO, 5% Mg:CCO, and 10% Mg:CCO HTL, respectively. For MAPbI$_3$ PSCs, 10, 11, and 10 devices were measured for CCO, 5% Mg:CCO, and 10% Mg:CCO HTL, respectively, under both forward and reverse scans.

Several groups have reported elemental diffusion from inorganic transport layer into MAPbI$_3$ when using CdS ETL [58], CrO$_x$ [59], and CuI [60] HTLs. In order to examine this possibility, we performed XPS studies on the surfaces of MAPbI$_3$ films on top of ITO/HTL. Figure 11 shows the normalized XPS spectra of (a) survey, (b) Cu 2p, (c) Cr 2p, and (d) Mg 2p core levels for MAPbI$_3$ films processed on top of CCO (black), 5% Mg:CCO (red), and 10% Mg:CCO (blue) HTLs. For all HTLs, all peaks in the survey spectra are indexed as the component elements (C, N, Pb, and I) of MAPbI$_3$, consistent with our previous result [27]. The Cu 2p and Cr 2p spectral ranges are free of any peaks corresponding to Cu 2p or Cr 2p (dashed lines), indicating no presence of Cu and Cr elements at the surface of MAPbI$_3$ layer. The Mg 2p spectrum shows two peaks at 48.2 eV and 46.6 eV corresponding to the binding energy of the I 4d orbitals [61]. No Mg 2p peak at 50.8 eV (dashed line) was detected, indicating no presence of Mg element. Thus, if metal diffusion from CCO and Mg:CCO HTLs into MAPbI$_3$ occurs, it does so at a level below the sensitivity of XPS. This result is consistent with thermodynamic calculation: the calculated formation enthalpy of CCO is -6.0 eV [22], significantly lower compared to that of CdS (−1.5 eV) [62] and CuI (−0.3 eV) [63] and slightly lower compared to that of Cr$_2$O$_3$ (−5.9 eV) [64]. Mg:CCO has the same crystalline structure as CCO and the Mg doping content is small, thus, the formation enthalpy of Mg:CCO is expected to be similar to that of CCO. Thus, CCO and Mg:CCO are more stable and less likely to decompose or react than the aforementioned HTLs. Nevertheless, additional experimentation is warranted to explore the possibility of reactivity between CCO/Mg:CCO and perovskite phases.

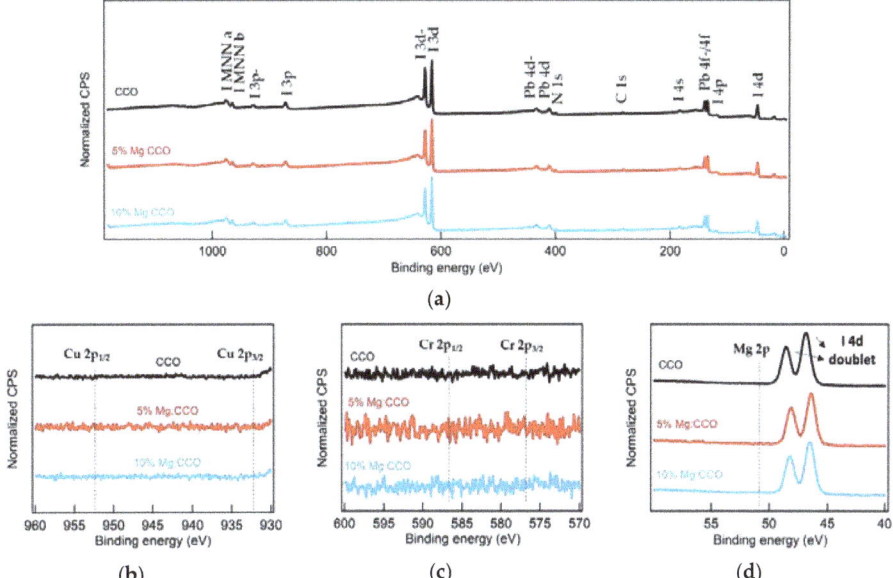

Figure 11. Normalized XPS (**a**) survey, (**b**) Cu 2p, (**c**) Cr 2p, and (**d**) Mg 2p spectra at the surfaces of MAPbI$_3$ films on top of ITO/HTL. HTLs are CCO (black), 5% Mg:CCO (red), and 10% Mg:CCO (blue). In (**a**), all peaks are indexed by the component elements (C, N, Pb, and I) of MAPbI$_3$. In (**b**–**d**), the dotted lines show binding energies for the Cu 2p, Cr 2p, and Mg 2p core levels of CCO and Mg:CCO. The positions of Cu 2p$_{1/2}$, Cu 2p$_{3/2}$, Cr 2p$_{1/2}$, and Cr 2p$_{3/2}$ peaks are indexed according to our previous CCO reports [24,27]. The position of Mg 2p peak is indexed according to the report from Hoogewijs et al. [65]. In (**d**), the peaks correspond to the I 4d orbitals; peaks due to the Mg 2p orbitals are not observed.

In order to explore charge transport at the CCO and Mg:CCO/MAPbI$_3$ interface, we performed TRPL measurements. Figure 12a shows the PL emission spectrum for ITO/MAPbI$_3$ (green), ITO/CCO/MAPbI$_3$ (black), ITO/5% Mg:CCO/MAPbI$_3$ (red), and ITO/10% Mg:CCO/MAPbI$_3$ (blue). For all samples, the main PL emission peaks are at ~750 nm, consistent with the literature [66]. PL intensities for MAPbI$_3$ on top of CCO and Mg:CCO HTLs are lower compared to that of MAPbI$_3$ on ITO, indicating CCO and Mg:CCO HTLs are effective in promoting charge transfer. However, PL intensities are similar among MAPbI$_3$ on top of CCO and Mg:CCO HTLs. Figure 12b shows the normalized TRPL decay kinetics for ITO/MAPbI$_3$ (green), ITO/CCO/MAPbI$_3$ (black), ITO/5% Mg:CCO/MAPbI$_3$ (red), and ITO/10% Mg:CCO/MAPbI$_3$ (blue). With the addition of CCO and Mg:CCO HTLs, a faster PL decay is observed relative to ITO/MAPbI$_3$. The inset table in Figure 12b shows the PL lifetimes extracted from three exponential fits in all samples (lines in Figure 12b). The τ_1, τ_2, and τ_3 lifetimes of the ITO/MAPbI$_3$ structure are 1.5 ns, 4.9 ns, and 16.3 ns, respectively. After adding CCO and Mg:CCO HTLs, the τ_1, τ_2, and τ_3 decreases to 0.7 ns, ~3.0 ns, and ~11.0 ns, respectively, indicating enhanced charge extraction and consistent with the literature result [66]. Again, there are no significant differences in PL lifetimes among films of MAPbI$_3$ on CCO and Mg:CCO HTLs. Mg doping in CCO is expected to lead lower PL intensity and shorter lifetimes. However, these effects are not discernable in our TRPL results, presumably because they may be confounded by factors besides charge transfer, such as surface recombination [67].

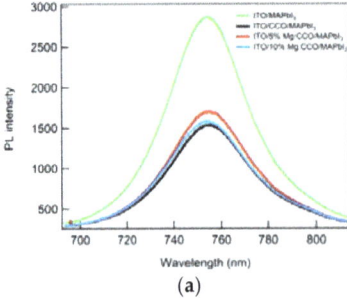 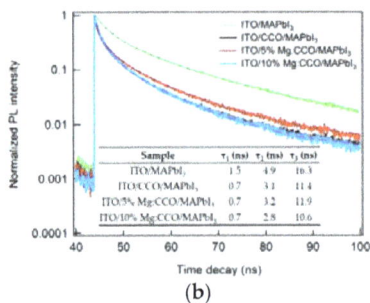

Figure 12. (**a**) Photoluminescence (PL) emission spectra and (**b**) time-resolved photoluminescence (TRPL) decay for ITO/MAPbI$_3$ (green), ITO/CCO/MAPbI$_3$ (black), ITO/5% Mg:CCO/MAPbI$_3$ (red), and ITO/10% Mg:CCO/MAPbI$_3$ (blue). In (**b**), the lines are fits to three exponential decays: dotted green line for ITO/MAPbI$_3$, solid grey line for ITO/CCO/MAPbI$_3$, dashed brown line for ITO/5% Mg:CCO/MAPbI$_3$, and dotted-dashed blue line for ITO/10% Mg:CCO/MAPbI$_3$. The inset table in (**b**) shows the fitted PL lifetimes for all samples.

Figure 13 shows the stabilized photocurrents and efficiencies for representative MAPbI$_3$ cells with CCO, 5% Mg:CCO, and 10% Mg:CCO HTLs. Time-dependent photocurrent measurements are taken at a bias of ~0.8 V with stabilized photocurrent values of 17.35 mA cm^{-2}, 18.05 mA cm^{-2}, and 17.81 mA cm^{-2} for CCO, 5% Mg:CCO, and 10% Mg:CCO HTLs, respectively. The values of the stabilized photocurrent are higher with Mg doping, reflecting the increases in J_{sc} and FF observed in forward and reverse scans for Mg:CCO MAPbI$_3$ PSCs. The stabilized efficiencies for CCO, 5% Mg:CCO, and 10% Mg:CCO HTLs are 13.89%, 14.43%, and 14.26%, respectively.

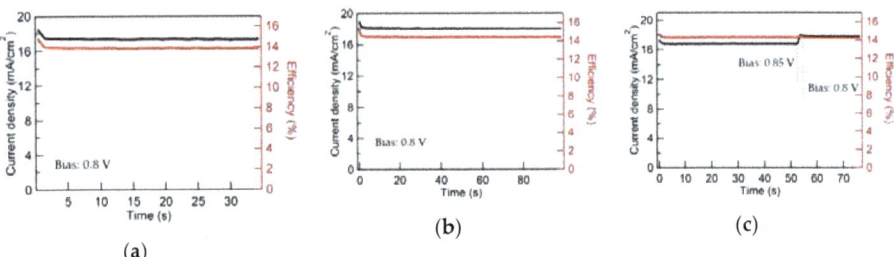

Figure 13. The stabilized photocurrents and efficiencies for the representative MAPbI$_3$ cells with (**a**) CCO, (**b**) 5% Mg:CCO, and (**c**) 10% Mg:CCO HTLs. In (**a**,**b**), the applied bias is at 0.8 V. In (**c**), the applied bias is initially at 0.85 V. After 50 s, it switches to 0.8 V.

Figure 14 shows the bar charts of average J_{sc} for P3HT:PC$_{61}$BM OSCs, PFBT2Se2Th:PC$_{71}$BM OSCs, PTB7-Th:ITIC OSCs, and MAPbI$_3$ PSCs under forward and reverse scans for CCO (black color), 5% Mg:CCO (red color), and 10% Mg:CCO (blue color) HTLs. The average J_{sc} of all OSCs and MAPbI$_3$ PSCs are higher with Mg:CCO HTLs. The small average J_{sc} increases in all systems may be partially attributed to the better conductivity of Mg:CCO HTLs resulting from the increased WF with respect to IE with Mg doping. Additionally, the broadband increase with Mg doping content in the EQE spectra of PSCs (Figure 10d inset) signifies that the increased HTL work function contributes to a stronger electric field within the device, more efficiently extracting photoexcited carriers regardless of the depth at which the generating photons are absorbed. If V_{oc} and FF are independent of Mg doping, the PCE may be expected to increase due to the boost in J_{sc}. However, they do not show a consistent trend from batch to batch. V_{oc} and FF are more susceptible to film roughness, which can vary due to aggregation of the nanoparticles in the suspensions and variation in spin coating conditions. The tradeoff between J_{sc} and V_{oc}/FF results in little or no statistical *PCE* improvement (Table 4).

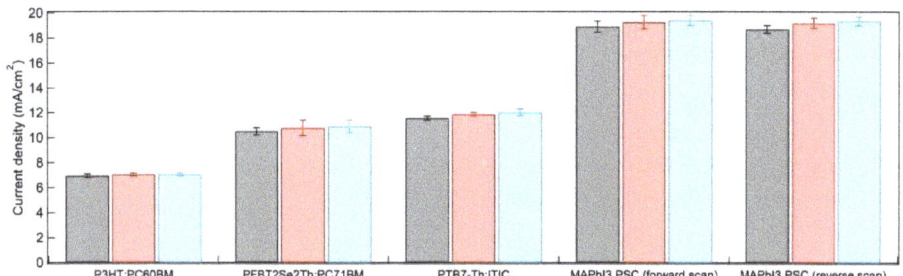

Figure 14. The average J_{sc} barcharts with error bars for P3HT:PC$_{61}$BM OSCs, PFBT2Se2Th:PC$_{71}$BM OSCs, PTB7-Th:ITIC OSCs, and MAPbI$_3$ PSCs under forward and reverse scans for CCO (black color), 5% Mg:CCO (red color), and 10% Mg:CCO (blue color) HTLs.

4. Conclusions

In summary, we synthesized CCO and Mg:CCO nanoparticles and successfully applied them as HTLs in OSCs and PSCs. Mg incorporation induces a slight lattice expansion by substituting larger ionic radii Mg^{2+} into the Cr^{3+} site. Rietveld refinement suggests that Mg doping decreases CCO nanoparticle size along the c axis but increases CCO nanoparticle size along the in-plane directions. Overall, both XRD and TEM results indicate that nanoparticle sizes are smaller with Mg doping. The average value of the direct E_g is (3.26 ± 0.03) eV in all nanoparticle films. The *WF* values for all Mg concentrations are larger than the *IE* values, and their difference (*WF − IE*) increases with Mg concentration, consistent with increased p-type conductivity reported in the literature. OSCs and PSCs based on Mg:CCO HTLs show a consistent increase in average J_{sc} in all four absorber systems despite large uncertainties; however, an overall enhancement in *PCE* is not clearly discernible (except in PSCs) due to different trends in other parameters and sample variation. No elemental (Cu, Cr, and Mg) diffusion from CCO and Mg:CCO HTLs is detected by XPS at the surface of MAPbI$_3$ films. CCO and Mg:CCO HTLs effectively extract charge from the absorber, as evident in more PL quenching and shorter lifetimes when MAPbI$_3$ is deposited on the HTLs. Mg doping in CCO HTLs enhances the stabilized efficiency for MAPbI$_3$ PSCs. This work provides new insights related to the role that an Mg:CCO HTL may play in improving performance in a wide range of OSCs and MAPbI$_3$ PSCs.

Author Contributions: Conceptualization, J.W.P.H.; Nanoparticle synthesis, making suspensions and thin films, XRD, SEM, EDX, DLS, UV-vis, PESA, Kelvin Probe, ellipsometry, B.Z.; TEM, XRD, XPS analyses, S.T. and B.Z.; OSC fabrication and PL sample preparation, B.Z. and W.X.; PSC fabrication and XPS, W.A.D.-S. and D.B.M.; Se polymer synthesis, F.-Y.C. and Y.-J.C.; PL quenching measurements and analyses, Y.Z. and A.V.M.; writing—original draft preparation, B.Z.; writing—review and editing, S.T., J.W.P.H., W.A.D.-S., and D.B.M.

Funding: The work done at University of Texas at Dallas (J.W.P.H., B.Z., W.X., S.T.) was sponsored by the National Science Foundation (NSF) (Grant No. DMR-1305893). PL spectroscopy work (Y.Z. and A.V.M.) was supported by NSF-CAREER grant #1350800. J.W.P.H. acknowledges the support from Texas Instruments Distinguished Chair in Nanoelectronics. The work done at Duke University was funded by the Office of Energy Efficiency and Renewable Energy (EERE), U.S. Department of Energy, under Award Number DE-EE0006712 (W.A.D.-S., D.B.M.). W.A.D.-S. acknowledges support from the Fitzpatrick Institute for Photonics John T. Chambers Scholarship.

Acknowledgments: We thank Trey Daunis for review and editing this manuscript and Ivanic Bojana for the help of PL quenching measurements.

Conflicts of Interest: The authors declare no conflict of interest.

References

1. Jung, E.H.; Jeon, N.J.; Park, E.Y.; Moon, C.S.; Shin, T.J.; Yang, T.-Y.; Noh, J.H.; Seo, J. Efficient, stable and scalable perovskite solar cells using poly(3-hexylthiophene). *Nature* **2019**, *567*, 511–515. [CrossRef]
2. Cui, Y.; Yao, H.; Zhang, J.; Zhang, T.; Wang, Y.; Hong, L.; Xian, K.; Xu, B.; Zhang, S.; Peng, J.; et al. Over 16% efficiency organic photovoltaic cells enabled by a chlorinated acceptor with increased open-circuit voltages. *Nat. Commun.* **2019**, *10*, 2515. [CrossRef] [PubMed]
3. Hany, R.; Lin, H.; Castro, F.A. Focus issue on organic and hybrid photovoltaics. *Sci. Technol. Adv. Mater.* **2019**, *20*, 42–43. [CrossRef] [PubMed]
4. Yip, H.-L.; Jen, A.K.-Y. Recent advances in solution-processed interfacial materials for efficient and stable polymer solar cells. *Energy Environ. Sci.* **2012**, *5*, 5994–6011. [CrossRef]
5. Yan, C.; Barlow, S.; Wang, Z.; Yan, H.; Jen, A.K.-Y.; Marder, S.R.; Zhan, X. Non-fullerene acceptors for organic solar cells. *Nat. Rev. Mater.* **2018**, *3*, 18003. [CrossRef]
6. Teh, C.H.; Daik, R.; Lim, E.L.; Yap, C.C.; Ibrahim, M.A.; Ludin, N.A.; Sopian, K.; Teridi, M.A.M. A review of organic small molecule-based hole-transporting materials for meso-structured organic–inorganic perovskite solar cells. *J. Mater. Chem. A* **2016**, *4*, 15788–15822. [CrossRef]
7. Su, Y.W.; Lan, S.C.; Wei, K.H. Organic photovoltaics. *Mater. Today* **2012**, *15*, 554–562. [CrossRef]
8. Liu, J.; Wu, Y.; Qin, C.; Yang, X.; Yasuda, T.; Islam, A.; Zhang, K.; Peng, W.; Chen, W.; Han, L. A dopant-free hole-transporting material for efficient and stable perovskite solar cells. *Energy Environ. Sci.* **2014**, *7*, 2963–2967. [CrossRef]
9. Kawano, K.; Pacios, R.; Poplavskyy, D.; Nelson, J.; Durrant, J.R. Degradation of organic solar cells due to air exposure. *Sol. Energy Mater. Sol. Cells* **2006**, *90*, 3520–3530. [CrossRef]
10. You, J.; Meng, L.; Song, T.-B.; Guo, T.-F.; Yang, Y.; Chang, W.-H.; Hong, Z.; Chen, H.; Zhou, H.; Chen, Q.; et al. Improved air stability of perovskite solar cells via solution-processed metal oxide transport layers. *Nat. Nanotechnol.* **2016**, *11*, 75–81. [CrossRef]
11. Stubhan, T.; Li, N.; Luechinger, N.A.; Halim, S.C.; Matt, G.J.; Brabec, C.J. High Fill Factor Polymer Solar Cells Incorporating a Low Temperature Solution Processed WO$_3$ Hole Extraction Layer. *Adv. Energy Mater.* **2012**, *2*, 1433–1438. [CrossRef]
12. Lee, Y.-J.; Yi, J.; Gao, G.F.; Koerner, H.; Park, K.; Wang, J.; Luo, K.; Vaia, R.A.; Hsu, J.W.P. Low-temperature solution-processed molybdenum oxide nanoparticle hole transport layers for organic photovoltaic devices. *Adv. Energy Mater.* **2012**, *2*, 1193–1197. [CrossRef]
13. Wang, J.; Xu, L.; Lee, Y.-J.; Villa, M.D.A.; Malko, A.V.; Hsu, J.W.P. Effects of Contact-Induced Doping on the Behaviors of Organic Photovoltaic Devices. *Nano Lett.* **2015**, *15*, 7627–7632. [CrossRef] [PubMed]
14. Hammond, S.R.; Meyer, J.; Widjonarko, N.E.; Ndione, P.F.; Sigdel, A.K.; Miedaner, A.; Lloyd, M.T.; Kahn, A.; Ginley, D.S.; Berry, J.J.; et al. Low-temperature, solution-processed molybdenum oxide hole-collection layer for organic photovoltaics. *J. Mater. Chem.* **2012**, *22*, 3249–3254. [CrossRef]
15. Singh, R.; Singh, P.K.; Bhattacharya, B.; Rhee, H.-W. Review of current progress in inorganic hole-transport materials for perovskite solar cells. *Appl. Mater. Today* **2019**, *14*, 175–200. [CrossRef]
16. Jiang, F.; Choy, W.C.H.; Li, X.; Cheng, J. Post-treatment-Free Solution-Processed Non-stoichiometric NiO$_x$ Nanoparticles for Efficient Hole-Transport Layers of Organic Optoelectronic Devices. *Adv. Mater.* **2015**, *27*, 2930–2937. [CrossRef] [PubMed]
17. Hüfner, S. Electronic structure of NiO and related 3d-transition-metal compounds. *Adv. Phys.* **1994**, *43*, 183–356. [CrossRef]
18. Renaud, A.; Chavillon, B.; Pleux, L.L.; Pellegrin, Y.; Blart, E.; Boujtita, M.; Pauporté, T.; Cario, L.; Jobic, S.; Odobel, F. CuGaO$_2$: A promising alternative for NiO in p-type dye solar cells. *J. Mater. Chem.* **2012**, *22*, 14353–14356. [CrossRef]
19. Kawazoe, H.; Yasukawa, M.; Hyodo, H.; Kurita, M.; Yanagi, H.; Hosono, H. P-type electrical conduction in transparent thin films of CuAlO$_2$. *Nature* **1997**, *389*, 939–942. [CrossRef]
20. Sheets, W.C.; Mugnier, E.; Barnabé, A.; Marks, T.J.; Poeppelmeier, K.R. Hydrothermal Synthesis of Delafossite-Type Oxides. *Chem. Mater.* **2006**, *18*, 7–20. [CrossRef]
21. Marquardt, M.A.; Ashmore, N.A.; Cann, D.P. Crystal chemistry and electrical properties of the delafossite structure. *Thin Solid Films* **2006**, *496*, 146–156. [CrossRef]

22. Scanlon, D.O.; Watson, G.W. Understanding the p-type defect chemistry of CuCrO$_2$. *J. Mater. Chem.* **2011**, *21*, 3655–3663. [CrossRef]
23. Arnold, T.; Payne, D.J.; Bourlange, A.; Hu, J.P.; Egdell, R.G.; Piper, L.F.J.; Colakerol, A.D.M.; Glans, P.A.; Learmonth, T.; Smith, K.E.; et al. X-ray spectroscopic study of the electronic structure of. *Phys. Rev. B* **2009**, *79*, 075102. [CrossRef]
24. Wang, J.; Lee, Y.-J.; Hsu, J.W.P. Sub-10 nm copper chromium oxide nanocrystals as a solution processed p-type hole transport layer for organic photovoltaics. *J. Mater. Chem. C* **2016**, *4*, 3607–3613. [CrossRef]
25. Xiong, D.; Xu, Z.; Zeng, X.; Zhang, W.; Chen, W.; Xu, X.; Wang, M.; Cheng, Y.-B. Hydrothermal synthesis of ultrasmall CuCrO$_2$ nanocrystal alternatives to NiO nanoparticles in efficient p-type dye-sensitized solar cells. *J. Mater. Chem.* **2012**, *22*, 24760–24768. [CrossRef]
26. Zhang, H.; Wang, H.; Zhu, H.; Chueh, C.-C.; Chen, W.; Yang, S.; Jen, A.K.-Y. Low-Temperature Solution-Processed CuCrO$_2$ Hole-Transporting Layer for Efficient and Photostable Perovskite Solar Cells. *Adv. Energy Mater.* **2018**, *8*, 1702762. [CrossRef]
27. Dunlap-Shohl, W.A.; Daunis, T.B.; Wang, X.; Wang, J.; Zhang, B.; Barrera, D.; Yan, Y.; Hsu, J.W.P.; Mitzi, D.B. Room-temperature fabrication of a delafossite CuCrO$_2$ hole transport layer for perovskite solar cells. *J. Mater. Chem. A* **2018**, *6*, 469–477. [CrossRef]
28. Yang, B.; Ouyang, D.; Huang, Z.; Ren, X.; Zhang, H.; Choy, W.C.H. Multifunctional Synthesis Approach of In:CuCrO$_2$ Nanoparticles for Hole Transport Layer in High-Performance Perovskite Solar Cells. *Adv. Funct. Mater.* **2019**, *29*, 1902600. [CrossRef]
29. Akin, S.; Liu, Y.; Dar, M.I.; Zakeeruddin, S.M.; Grätzel, M.; Turan, S.; Sonmezoglu, S. Hydrothermally processed CuCrO$_2$ nanoparticles as an inorganic hole transporting material for low-cost perovskite solar cells with superior stability. *J. Mater. Chem. A* **2018**, *6*, 20327–20337. [CrossRef]
30. Zheng, S.Y.; Jiang, G.S.; Su, J.R.; Zhu, C.F. The structural and electrical property of CuCr$_{1-x}$Ni$_x$O$_2$ delafossite compounds. *Mater. Lett.* **2006**, *60*, 3871–3873. [CrossRef]
31. Nagarajan, R.; Draeseke, A.D.; Sleight, A.W.; Tate, J. P-type conductivity in CuCr$_{1-x}$Mg$_x$O$_2$ films and powders. *J. Appl. Phys.* **2001**, *89*, 8022–8025. [CrossRef]
32. Asemi, M.; Ghanaatshoar, M. Conductivity improvement of CuCrO$_2$ nanoparticles by Zn doping and their application in solid-state dye-sensitized solar cells. *Ceram. Int.* **2016**, *42*, 6664–6672. [CrossRef]
33. Schiavo, E.; Latouche, C.; Barone, V.; Crescenzi, O.; Pavone, M. An ab initio study of Cu-based delafossites as an alternative to nickel oxide in photocathodes: Effects of Mg-doping and surface electronic features. *Phys. Chem. Chem. Phys.* **2018**, *20*, 14082. [CrossRef] [PubMed]
34. Fang, Z.-J.; Zhu, J.-Z.; Zhou, J.; Mo, M. Defect properties of CuCrO$_2$: A density functional theory calculation. *Chin. Phys. B* **2012**, *21*, 087105. [CrossRef]
35. Bywalez, R.; Götzendörfer, S.; Löbmann, P. Structural and physical effects of Mg-doping on p-type CuCrO$_2$ and CuAl$_{0.5}$Cr$_{0.5}$O$_2$ thin films. *J. Mater. Chem.* **2010**, *20*, 6562–6570. [CrossRef]
36. Kaya, İ.C.; Sevindik, M.A.; Akyıldız, H. Characteristics of Fe- and Mg-doped CuCrO$_2$ nanocrystals prepared by hydrothermal synthesis. *J. Mater. Sci. Mater. Electron.* **2016**, *27*, 2404–2411. [CrossRef]
37. Monteiro, J.F.H.L.; Monteiro, F.C.; Jurelo, A.R.; Mosca, D.H. Conductivity in (Ag, Mg)-doped delafossite oxide CuCrO$_2$. *Ceram. Int.* **2018**, *44*, 14101–14107. [CrossRef]
38. Asemi, M.; Ghanaatshoar, M. Influence of TiO$_2$ particle size and conductivity of the CuCrO$_2$ nanoparticles on the performance of solid-state dye-sensitized solar cells. *Bull. Mater. Sci.* **2017**, *40*, 1379–1388. [CrossRef]
39. Xiong, D.; Zhang, W.; Zeng, X.; Xu, Z.; Chen, W.; Cui, J.; Wang, M.; Sun, L.; Cheng, Y.-B. Enhanced Performance of p-Type Dye-Sensitized Solar Cells Based on Ultrasmall Mg-Doped CuCrO$_2$ Nanocrystals. *ChemSusChem* **2013**, *6*, 1432–1437. [CrossRef]
40. Jeong, S.; Seo, S.; Shin, H. p-Type CuCrO$_2$ particulate films as the hole transporting layer for CH$_3$NH$_3$PbI$_3$ perovskite solar cells. *RSC Adv.* **2018**, *8*, 27956–27962. [CrossRef]
41. Cao, F.-Y.; Tseng, C.-C.; Lin, F.-Y.; Chen, Y.; Yan, H.; Cheng, Y.-J. Selenophene-Incorporated Quaterchalcogenophene-Based Donor–Acceptor Copolymers to Achieve Efficient Solar Cells with J$_{sc}$ Exceeding 20 mA/cm^2. *Chem. Mater.* **2017**, *29*, 10045–10052. [CrossRef]
42. Ahn, N.; Son, D.-Y.; Jang, I.-H.; Kang, S.M.; Choi, M.; Park, N.-G. Highly Reproducible Perovskite Solar Cells with Average Efficiency of 18.3% and Best Efficiency of 19.7% Fabricated via Lewis Base Adduct of Lead(II) Iodide. *J. Am. Chem. Soc.* **2015**, *137*, 8696–8699. [CrossRef] [PubMed]

43. Nguyen, H.M.; Seitz, O.; Peng, W.; Gartstein, Y.N.; Chabal, Y.J.; Malko, A.V. Efficient Radiative and Nonradiative Energy Transfer from Proximal CdSe/ZnS Nanocrystals into Silicon Nanomembranes. *ACS Nano* **2012**, *6*, 5574–5582. [CrossRef] [PubMed]
44. Toby, B.H. R factors in Rietveld analysis: How good is good enough? *Powder Diffr.* **2006**, *21*, 67–70. [CrossRef]
45. Jansen, E.; Schäfer, W.; Will, G. R values in analysis of powder diffraction data using Rietveld refinement. *J. Appl. Cryst.* **1994**, *27*, 492–496. [CrossRef]
46. Boekestein, A.; Stadhouders, A.M.; Stols, A.L.H.; Roomans, G.M. A comparison of ZAF-correction methods in quantitative X-ray microanalysis of light-element specimens. *Ultramicroscopy* **1983**, *12*, 65–68. [CrossRef]
47. Miclau, M.; Ursu, D.; Kumar, S.; Grozescu, I. Hexagonal polytype of $CuCrO_2$ nanocrystals obtained by hydrothermal method. *J. Nanopart. Res.* **2012**, *14*, 1110. [CrossRef]
48. Beverskog, B.; Puigdomenech, I. Revised pourbaix diagrams for chromium at 25–300 °C. *Corros. Sci.* **1997**, *39*, 43–57. [CrossRef]
49. Yu, S.-H.; Yoshimura, M. Direct fabrication of ferrite MFe_2O_4 (M = Zn, Mg)/Fe composite thin films by soft solution processing. *Chem. Mater.* **2000**, *12*, 3805–3810. [CrossRef]
50. Beverskog, B.; Puigdomenech, I. Revised pourbaix diagrams for copper at 25 to 300°C. *J. Electrochem. Soc.* **1997**, *144*, 3476–3483. [CrossRef]
51. Jiang, J.; Oberdörster, G.; Biswas, P. Characterization of size, surface charge, and agglomeration state of nanoparticle dispersions for toxicological studies. *J. Nanopart. Res.* **2009**, *11*, 77–89. [CrossRef]
52. Buford, M.C.; Hamilton, R.F.; Holian, A. A comparison of dispersing media for various engineered carbon nanoparticles. *Part. Fibre Toxicol.* **2007**, *4*, 6. [CrossRef] [PubMed]
53. Thampy, S.; Ibarra, V.; Lee, Y.-J.; McCool, G.; Cho, K.; Hsu, J.W.P. Effects of synthesis conditions on structure and surface properties of $SmMn_2O_5$ mullite-type oxide. *Appl. Surf. Sci.* **2016**, *385*, 490–497. [CrossRef]
54. Cano, E.; Torres, C.L.; Bastidas, J.M. An XPS study of copper corrosion originated by formic acid vapour at 40% and 80% relative humidity. *Mater. Corros.* **2001**, *52*, 667–676. [CrossRef]
55. Ursu, D.; Miclau, M. Thermal stability of nanocrystalline 3R-$CuCrO_2$. *J. Nanopart. Res.* **2014**, *16*, 2160. [CrossRef]
56. Biesinger, M.C.; Payne, B.P.; Grosvenor, A.P.; Lau, L.W.M.; Gerson, A.R.; Smart, R.S.C. Resolving surface chemical states in XPS analysis of first row transition metals, oxides and hydroxides: Cr, Mn, Fe, Co and Ni. *Appl. Surf. Sci.* **2011**, *257*, 2717–2730. [CrossRef]
57. Ratcliff, E.L.; Meyer, J.; Steirer, K.X.; Garcia, A.; Berry, J.J.; Ginley, D.S.; Olson, D.C.; Kahn, A.; Armstrong, N.R. Evidence for near-Surface NiOOH Species in Solution-Processed NiO_x Selective Interlayer Materials: Impact on Energetics and the Performance of Polymer Bulk Heterojunction Photovoltaics. *Chem. Mater.* **2011**, *23*, 4988–5000. [CrossRef]
58. Dunlap-Shohl, W.A.; Younts, R.; Gautam, B.; Gundogdu, K.; Mitzi, D.B. Effects of Cd Diffusion and Doping in High-Performance Perovskite Solar Cells Using CdS as Electron Transport Layer. *J. Phys. Chem. C* **2016**, *120*, 16437–16445. [CrossRef]
59. Qin, P.; He, Q.; Yang, G.; Yu, X.; Xiong, L.; Fang, G. Metal ions diffusion at heterojunction chromium Oxide/$CH_3NH_3PbI_3$ interface on the stability of perovskite solar cells. *Surf. Interfaces* **2018**, *10*, 93–99. [CrossRef]
60. Khadka, D.B.; Shirai, Y.; Yanagida, M.; Miyano, K. Unraveling the Impacts Induced by Organic and Inorganic Hole Transport Layers in Inverted Halide Perovskite Solar Cells. *ACS Appl. Mater. Interfaces* **2019**, *11*, 7055–7065. [CrossRef]
61. Hoste, S.; Van De Vondel, D.F.; Van Der Kelen, G.P. XPS Spectra of organometallic phenyl compounds of P, As, Sb and Bi. *J. Electron Spectrosc. Relat. Phenom.* **1979**, *17*, 191–195. [CrossRef]
62. Xu, F.; Ma, X.; Kauzlarich, S.M.; Navrotsky, A. Enthalpies of formation of CdS_xSe_{1-x} solid solutions. *J. Mater. Res.* **2009**, *24*, 1368–1374. [CrossRef]
63. Materials Project. Available online: https://materialsproject.org/#apps/reactioncalculator/\protect\T1\textbraceleft%22reactants%22%3A[%22Cu%22%2C%22I%22]%2C%22products%22%3A[%22CuI%22]\protect\T1\textbraceright (accessed on 7 September 2019).
64. Materials Project. Available online: https://materialsproject.org/#apps/reactioncalculator/\protect\T1\textbraceleft%22reactants%22%3A[%22Cr%22%2C%22O%22]%2C%22products%22%3A[%22Cr2O3%22]\protect\T1\textbraceright (accessed on 7 September 2019).

65. Hoogewijs, R.; Fiermans, L.; Vennik, J. Electronic relaxation processes in the KLL' auger spectra of the free magnesium atom, solid magnesium and MgO. *J. Electron Spectrosc. Relat. Phenom.* **1977**, *11*, 171–183. [CrossRef]
66. Khadka, D.B.; Shirai, Y.; Yanagida, M.; Ryan, J.W.; Miyano, K. Exploring the effects of interfacial carrier transport layers on device performance and optoelectronic properties of planar perovskite solar cells. *J. Mater. Chem. C* **2017**, *5*, 8819–8827. [CrossRef]
67. Wang, J.; Fu, W.; Jariwala, S.; Sinha, I.; Jen, A.K.-Y.; Ginger, D.S. Reducing Surface Recombination Velocities at the Electrical Contacts Will Improve Perovskite Photovoltaics. *ACS Energy Lett.* **2019**, *4*, 222–227. [CrossRef]

© 2019 by the authors. Licensee MDPI, Basel, Switzerland. This article is an open access article distributed under the terms and conditions of the Creative Commons Attribution (CC BY) license (http://creativecommons.org/licenses/by/4.0/).

Article

High-Sensitive Ammonia Sensors Based on Tin Monoxide Nanoshells

Han Wu [1,2], Zhong Ma [1,2], Zixia Lin [3], Haizeng Song [1,2], Shancheng Yan [4,*] and Yi Shi [1,2,*]

1. Collaborative Innovation Center of Advanced Microstructures, Nanjing University, Nanjing 210093, China; wuhan1106@yeah.net (H.W.); mazhongnj@163.com (Z.M.); songhaizeng0501@foxmail.com (H.S.)
2. National Laboratory of Solid State Microstructures, School of Electronic Science and Engineering, Nanjing University, Nanjing 210093, China
3. Testing center, Yangzhou University, Yangzhou 225009, China; linzixia@hotmail.com
4. School of Geography and Biological Information, Nanjing University of Posts and Telecommunications, Nanjing 210023, China
* Correspondence: yansc@njupt.edu.cn (S.Y.); yshi@nju.edu.cn (Y.S.); Tel.: +86-25-85866634 (S.Y.); +86-25-83621220 (Y.S.)

Received: 16 January 2019; Accepted: 20 February 2019; Published: 7 March 2019

Abstract: Ammonia (NH_3) is a harmful gas contaminant that is part of the nitrogen cycle in our daily lives. Therefore, highly sensitive ammonia sensors are important for environmental protection and human health. However, it is difficult to detect low concentrations of ammonia ($\leq$50 ppm) using conventional means at room temperature. Tin monoxide (SnO), a member of IV–VI metal monoxides, has attracted much attention due to its low cost, environmental-friendly nature, and higher stability compared with other non-oxide ammonia sensing material like alkaline metal or polymer, which made this material an ideal alternative for ammonia sensor applications. In this work, we fabricated high-sensitive ammonia sensors based on self-assembly SnO nanoshells via a solution method and annealing under 300 °C for 1 h. The as fabricated sensors exhibited the response of 313%, 874%, 2757%, 3116%, and 3757% ($\Delta G/G$) under ammonia concentration of 5 ppm, 20 ppm, 50 ppm, 100 ppm, and 200 ppm, respectively. The structure of the nanoshells, which have curved shells that provide shelters for the core and also possess a large surface area, is able to absorb more ammonia molecules, leading to further improvements in the sensitivity. Further, the SnO nanoshells have higher oxygen vacancy densities compared with other metal oxide ammonia sensing materials, enabling it to have higher performance. Additionally, the selectivity of ammonia sensors is also outstanding. We hope this work will provide a reference for the study of similar structures and applications of IV–VI metal monoxides in the gas sensor field.

Keywords: tin monoxide; nanoshell; ammonia sensor; solution method

---

1. Introduction

Recently, ammonia (NH_3) sensors have been widely studied by researchers and widely used in the high volume control of combustibles in the chemical industry, the control of emission of vehicles, and the monitoring of dairy products in the food industry. Currently, many nanomaterials have been utilized in ammonia sensors. These include two-dimension materials [1–5], IV–VI metal chalcogenides, conductive polymers, and alkali metal materials. Among these materials, owing to the low cost, high sensitivity, and environmentally-friendly nature, tin monoxide and tin dioxide have been used to fabricate ammonia gas sensors [6–16]. Compared with tin dioxide, tin monoxide synthesized under lower temperature is more stable and absorbs ammonia more easily when utilized as gas sensors [17,18].

As one of the common pollutants and toxic gases, ammonia (NH_3) can cause several effects on the human body like irritation of the eyes, skin, throat, and respiratory system. According to the US Occupational Safety and Health Administration (OSHA), the exposure of under 35 ppm of ammonia by volume in environmental air for 15 min or under 25 ppm of volume for 8 h potentially harms people's health [12,13]. However, it is impossible for humans to detect ammonia below 50 ppm, which reflects the importance of ammonia sensing. Hence, a highly sensitive and selective room temperature NH_3 gas sensor is highly desirable in today's world [19,20]. In previous work, the effect of material structure on the sensitivity of indium oxide-based ammonia sensors has been discussed [21]. Deren Yang et al. demonstrated that broken indium oxide nanotube structure with ultrahigh surface-to-volume ratio exhibited higher performance than regular nanotube, nanowire, and nanoparticle [21]. This is because the ultrahigh surface-to-volume ratio material can potentially provide larger interface to absorb gas [22]. Shell structure is another structure with ultrahigh surface-to-volume ratio that is suitable for gas sensing [23]. After referring to this work and the fabrication of CdS nanoshell structure [24], we synthesized a SnO nanoshell structure that also possesses high surface-to-volume ratio with the aim of improving the sensitivity of our ammonia sensors. SnO is a kind of monoxide and the mechanism of ammonia sensing is related to the redox reactions. As is similar to other metal oxide sensors, when the SnO is exposed to air, oxygen will be adsorbed on its surface, and oxygen molecules attract electrons. As a result, the conductivity of the SnO decreases. Then, when the sensor is exposed to a reducing gas such as NH_3, the reducing gas may react with the adsorbed oxygen molecules and release electrons into the SnO, thereby increasing the conductivity. From this mechanism, the oxygen adsorption in the primary step is very important for the performance of the sensor [14,21]. The oxygen adsorption relies on the oxygen vacancy of the material and high oxygen vacancy density of SnO nanoshell also contributes to its high sensitivity.

In this paper, we prepared $Sn_6O_4(OH)_4$ as precursors through a facile solution method and further prepared SnO nanoshell through different annealing conditions. The morphology, structure, and chemical composition of our samples were investigated by instruments. Among all samples annealed under different conditions, Sample 3 showed shell structure and the highest response in ammonia sensing. Compared with reported works, gas sensors prepared by Sample 3 showed a much higher response. These prepared sensors also showed outstanding selectivity and stability. The mechanism of response was revealed, and two factors that contributed to the high sensitivity of the as prepared sensor were ultrahigh surface-to-volume ratio and high oxygen vacancy density. This work provided novel structure for conductive materials which were suitable for a high performance gas sensor. We hope this work will provide new ideas for applications of IV–VI metal monoxides in the gas sensor field.

2. Experimental Details

2.1. Materials

In this experiment, all chemicals used were of analytical grade and were applied as-received, without further purification. Thioacetamide and NaOH powder were purchased from Sinopharm Chemical Reagent Co., Ltd (Shanghai, China). Stannous chloride ($SnCl_2 \cdot 2H_2O$) was purchased from Aladdin Industrial Corporation (Shanghai, China). Ultrapure water that was used in the experiment was purified using the Millipore water purification system (Millipore Corporation, Burlington, MA, USA). N-menthylpyrrolidinone (1-methyl-2-pyrrolidinone) (NMP) was purchased from Sigma-Aldrich (St. Louis, MO, USA).

2.2. Synthesis Methods

In this experiment, 45 mL of deionized water and 33.75 mg of thioacetamide were introduced into a 100 mL bottle. After stirring for 1 min, 30 mg of NaOH was added to the bottle. While the solution was stirred, 101.54 mg of $SnCl_2 \cdot 2H_2O$ was added and the color of the solution became milky white. The reaction of

$$SnCl_2 \cdot 2H_2O + H_2O \rightarrow Sn(OH)Cl + HCl \tag{1}$$

took place [25]. Afterwards, the solution was stirred for another 20 min. The suspension was centrifuged for 20 min at 10,000 rpm and the precipitate was dried at 80 °C for 12 h in a vacuum oven. From our previous study, dried precipitate was $Sn_6O_4(OH)_4$ which was a precursor of SnO nanoshell. The precursor showed shell structure and this structure could be kept only annealed under suitable conditions. In fact, we found only Sample 3 and 5 kept shell structure in subsequent tests and Sample 3 had higher density of shell than Sample 5. The dried precipitate was then annealed under different conditions. Under specific annealing conditions, the decomposition reaction of precursor

$$Sn_6O_4(OH)_4 \rightarrow 6SnO + H_2O \tag{2}$$

took place. In this work, Sample 1, 2, 3, and 4 were obtained after heating from 0 °C to 200 °C, 250 °C, 300 °C, and 350 °C for 1 h and kept at 200 °C, 250 °C, 300 °C, and 350 °C for 1 h, respectively. Sample 5 was obtained after heating from 0 °C to 350 °C for 30 min and kept at 350 °C for 30 min. In Supplementary Materials, we have added a flow chart to demonstrate the synthesis process in Figure S1. To demonstrate the difference between each sample from preparing to investigating, we concluded Table S1 in Support Information.

2.3. Characterization of Material

Field emission scanning electron microscopy (FESEM, JSM-7000 F, JEOL Ltd., Tokyo, Japan) was used to determine the morphology of the samples. Transmission electron microscopy (TEM) and high-resolution transmission electron microscopy (HRTEM) images were obtained using a TEM (FEI Tecnai G2 F30 S-Twin TEM, Georgia Tech, Atlanta, GA, USA) instrument. Raman spectrum was obtained using a Raman spectrometer (LabRamHR800, HORIBA, Ltd., Kyoto, Japan) that was excited by an Ar laser at 514.5 nm under 500 µW. The crystal phase properties of the samples were analyzed by a Bruker D8 Advance X-ray diffractometer (Bruker, Billerica, MA, USA) with Ni-filtered Cu Kα radiation at 40 kV and 40 mA and 2θ from 10° to 60° with a scan rate of 0.02°. X-ray photoelectron spectroscopy (XPS) analysis (PHI5000Versaprobe, ULVAC-PHI, Inc. Chigasaki, Kanagawa, Japan) was used to determine the chemical composition of the products.

2.4. Fabrication and Measurement of the Ammonia Sensor

To fabricate ammonia sensors, 20 mg sample were mixed with 40 µL NMP (N-menthylpyrrolidinone 1-methyl-2-pyrrolidinone) and were grinded in a mortar for 20 min. Through a paint pen, the mixture was coated on sensors purchased from Winsen Electronic Technology Co., Ltd. (Zhengzhou, Henan Province, China). Finally, sensors were welded on substrates. The configuration of the as fabricated device is shown in Figure 1a. The as fabricated sensor devices were dried for 20 min at 50 °C in a vacuum oven. Then, ammonia gas sensing was tested by Navigation 4000 Series Smart Sensor Tester purchased from Beijing ZhongKe Micro-Nano Networking Science Technology Co., Ltd. (Beijing, China). The sensor response S was measured by systematically exposing sensors to different concentrations of ammonia (5 ppm to 200 ppm) at room temperature using the following equation $S = (G_g - G_a)/G_a \times 100$, while G_a and G_g are the conductance of the sample in air and ammonia gas, respectively.

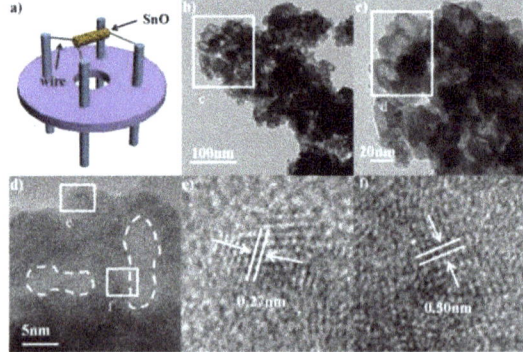

Figure 1. Configuration of the as fabricated devices and morphology of as prepared Sample 3. (a) The configuration of sensor devices. (b,c) TEM (transmission electron microscopy) image of Sample 3 at different magnifications. (d) TEM image of Sample 3 which expands the top-left section of (b). (e,f) HRTEM (high-resolution transmission electron microscopy) image of corresponding white solid frames marked in (c).

3. Results and Discussion

3.1. Characteristic of SnO Nanoshell Material

In this work, Sample 3, annealed and heated from 0 °C to 300 °C for 1 h and then kept under 300 °C for 1 h, possesses the nanoshell structure and exhibits the highest performance when utilized as the conductive material of ammonia sensors. The configuration of sensor devices and morphology of the as-synthesized Sample 3 is shown in Figure 1. Figure 1a shows the configuration of the sensor devices. Figure 1b shows the TEM image of Sample 3 which exhibits the accumulation of SnO nanoshells. Figure 1c shows the higher magnification TEM image of the top-left section of Figure 1b. We clearly observed the shell structure in the top-left region of Figure 1c. Further expansion of the white solid frame in Figure 1c shows the nanoshell structure with a hollow core, presented in Figure 1d, marked with white dashed lines. The morphology of Sample 3 in Figure 1 confirmed the existence of the nanoshell structure. The HRTEM image of the regions within the white solid frames of Figure 1d were shown in Figure 1e,f, showing a lattice fringe spacing of 0.27 nm and 0.30 nm, which corresponds to the (110) plane and (101) plane of SnO, respectively [10].

The crystal structure of the as-prepared Sample 3 is shown in Figure 2. Figure 2a shows the X-ray diffraction of Sample 2. Peaks at 18.2°, 29.8°, 33.2°, 37.1°, and 47.7° correspond to the (001), (101), (110), (002), and (200) crystal planes of SnO, respectively. All the diffraction peaks can be assigned to SnO (JCPDS Card No. 06-0395) [10,13]. From the X-ray diffraction, the intensity of (110) and (101) peaks was much higher than the other which matched well with these two planes and could be found easily in TEM. In addition, the Raman spectrum of Sample 3 is shown in Figure 2b. The peaks at 112 cm^{-1} and 210 cm^{-1} correspond to the B_{1g} and A_{1g} vibration mode of SnO, respectively [26]. In A_{1g} mode, Sn atoms vibrate towards or away from O atoms. B_{1g} mode corresponds to the out-of-plane vibrations of O atoms [27]. These characterizations confirmed that this material is SnO. The morphology and Raman spectra of other samples (Sample 1, Sample 2, Sample 4, and Sample 5) are provided in Figures S2–S5. From the supplementary material, Sample 3 and sample 5 showed shell structure and the density of shell structure of Sample 3 is higher than Sample 5.

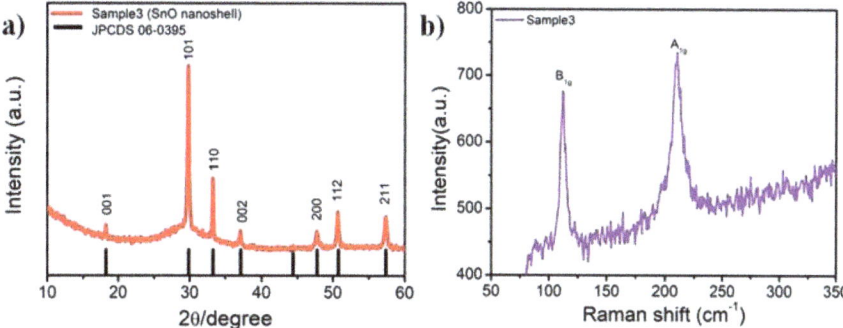

Figure 2. Characterization of the crystal structure of Sample 3. (**a**) X-ray diffraction of Sample 3; (**b**) Raman spectrum of Sample 3.

X-ray photoelectron spectroscopy shown in Figure 3 was used to analyze the surface chemical composition of Sample 3. As shown in Figure 3a, only peaks that correspond to Sn, O, and C were observed. Figure 3b shows the high resolution spectrum of Sn 3d. Peaks at 485.4 eV and 493.8 eV correspond to the energies of Sn $3d_{5/2}$ and Sn $3d_{3/2}$, respectively. The energy gap that reveals the amount of energy splitting between the two core levels is 8.38 eV, which corresponds to the energy splitting of Sn^{2+} [26]. Figure 3c shows the peak that corresponds to the Sn–O bond at 529.4 eV in O 1s spectroscopy. This result is consistent with previous work [28,29].

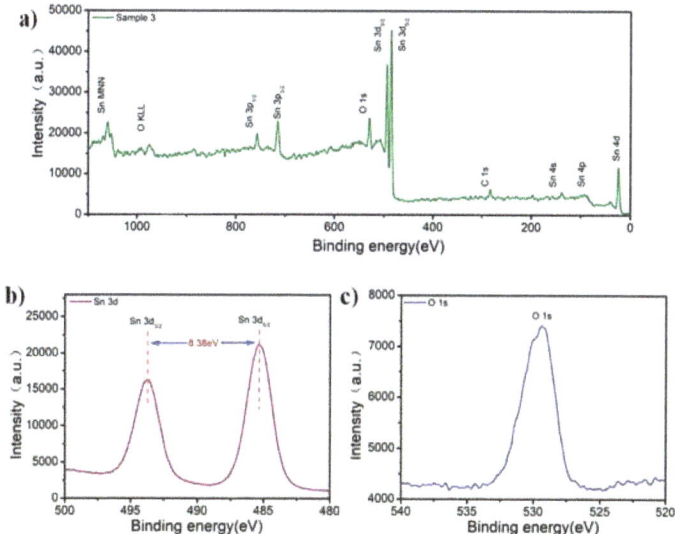

Figure 3. X-ray photoelectron spectroscopy of as-prepared Sample 3. (**a**) Sample 3 has Sn, O in the full spectrum; (**b**) High-resolution XPS of Sn 3d; (**c**) Sn–O bond in O 1s spectroscopy.

3.2. Test of Gas Sensor Device

In the experiment, we prepared 15 samples (three samples for each kind) for ammonia sensing tests from 0 to 200 ppm. A sketch added in supplementary material Figure S6 was shown to describe the mechanism. The response of the fabricated ammonia gas sensors, shown in Figure 4, reveals the huge difference between Sample 3 and other samples. As shown in Figure 4a,b, the response of Sample 3 is 313%, 874%, 2757%, 3116%, and 3757% under gas concentration of 5 ppm, 20 ppm, 50 ppm,

100 ppm, and 200 ppm, respectively, which is much higher than the other four samples. This result demonstrates that the large surface area of the nanoshell structure is able to absorb more ammonia. The response to different ammonia concentrations is shown in Figure 4c,d. From Figure 4c,d, all sensors fabricated had approximate linear response to ammonia below 20 ppm. Due to saturation of absorbance to ammonia, curves have a lower slope after 50 ppm and only the response of Sample 3 and Sample 2 maintain an increasing trend. The accurate response of Sample 3 from 0 to 40 ppm was shown in Figure S7. This test aimed to reveal that the sensor fabricated with Sample 3 is sufficiently sensitive to work under lower concentrations. In supplementary material, Table S1 was also used to compare the differences of each sample. From the table, we concluded that the main factor that contributed to the highest sensitivity of Sample 3 was its highest surface to volume ratio resulting from its highest density of shell structure.

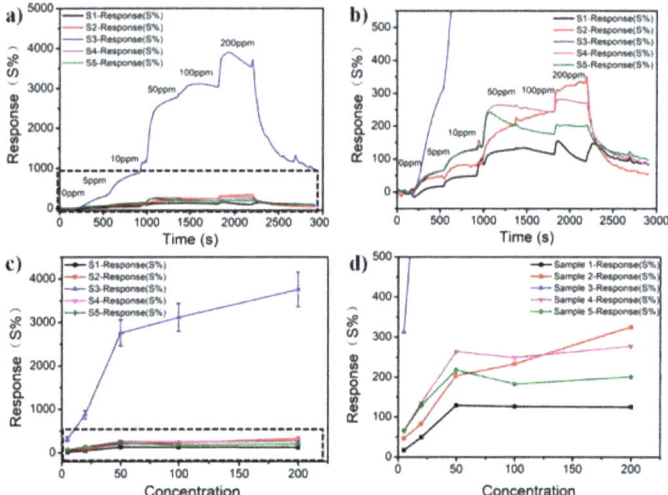

Figure 4. Response of ammonia gas sensors using different samples. (**a**) Response-recovery curves of the sensors up to 0–200 ppm NH$_3$; (**b**) Magnification of the black dashed pane in (**a**); (**c**) Response towards five (5–200 ppm) different concentrations of NH$_3$ in air; (**d**) Magnification of the black dashed pane in (**c**).

Response of as fabricated sensors compared with other work mentioned in this paper was shown in Figure 5a. Results in this figure were normalized as S = (R$_a$ − R$_g$)/R$_a$ × 100 where R$_a$ and R$_g$ are the resistance of the sample in air and ammonia gas, respectively. From the histogram, the response of our sensor is 97%, which is much higher than the other sensing materials reported in the literature [11–13,15,16]. The mechanism of ammonia sensing is related to the redox reactions. When the SnO is exposed to air, oxygen will be adsorbed on its surface, and oxygen molecules attract electrons. As a result, the conductivity of the SnO decreases. Then, when the sensor is exposed to a reducing gas such as NH$_3$, the reducing gas may react with the adsorbed oxygen molecules and release electrons into the SnO, thereby increasing the conductivity. During the sensing process, these reactions would take place:

$$O_2 \text{ (gas)} \rightarrow 2O \text{ (adsorbed)} \tag{3}$$

$$O \text{ (adsorbed)} + e^- \text{ (from SnO)} \rightarrow O^- \tag{4}$$

$$2 NH_3 \text{ (adsorbed)} + 3 O^- \rightarrow N_2 + 3 H_2O + 3e^- \tag{5}$$

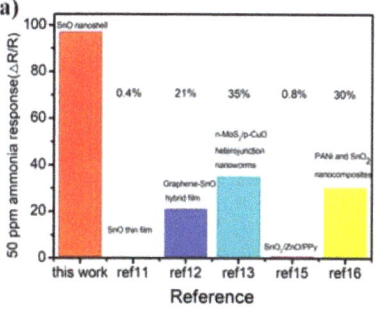 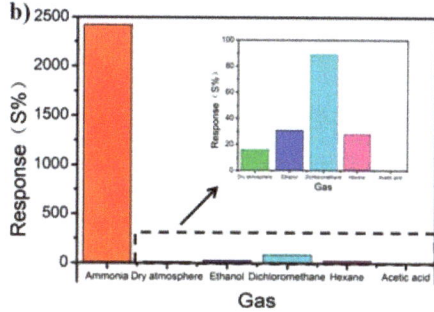

Figure 5. Response of as fabricated sensors compared with other work and in different gases. (a) Response of ammonia sensors compared with other work based on metal oxides. (The result was normalized as S = ($R_a - R_g$)/R_a × 100) (b) Response of as fabricated sensors in different gas environments.

The oxygen adsorption relies on the oxygen vacancy of the material [21]. As prepared SnO is an unsaturated metal oxide that tends to absorb oxygen and be further oxidized to SnO_2. From our previous work, the photoluminescence of oxygen vacancy on SnO nanoshell was studied which shows the high oxygen vacancy density of our SnO material [30]. In summary, the high response of our sensors depends on the high oxygen adsorption with high surface-to-volume ratio and high oxygen vacancy density of the as-prepared SnO nanoshell. Additionally, we investigated the test of selectivity of our sensors. Since there are many papers about volatile organic compound sensors based on metal oxide, we used different volatile organic compounds for comparison [31–34]. In Figure 5b, the response of our sensors in ammonia is also much higher than that of dry atmosphere or other organic gases. In the experiment, we used single gas for each test. This confirms the outstanding selectivity of our sensors. These results show the high performance of our samples. In the supplementary materials, the response (98 s) and recovery time (30 s) are shown in Figure S8. Comparison of response time and recovery time with previous work is shown in Table S2. These results show the high performance of our samples. Furthermore, we investigated the repeatability of the SnO nanoshell materials in a certain amount of 20 ppm NH_3 and found that the SnO nanoshell material possesses good repeatability for at least one week.

4. Conclusions

In conclusion, we prepared SnO nanoshell through a solution method and annealing. SEM, TEM, XRD, XPS, and Raman measurements were used to characterize the present samples. The SnO nanoshell exhibited high responses of 313%, 874%, 2757%, 3116%, and 3757% under gas concentrations of 5 ppm, 20 ppm, 50 ppm, 100 ppm, and 200 ppm, respectively. The mechanism of ammonia sensing is related to the redox reactions. From the mechanism, we realized high sensitivity was due to a large surface area and higher oxygen vacancy of the SnO nanoshell. This material also showed good selectivity and repeatability in ammonia sensing. This work can potentially aid in the study of similar structures and applications of IV–VI metal monoxides for real field applications.

Supplementary Materials: The following are available online at http://www.mdpi.com/2079-4991/9/3/388/s1, Figure S1: schematic figure about the synthesis process of tin mono oxide, Figure S2: characteristic of sample 1, Figure S3: characteristic of sample 2, Figure S4: characteristic of sample 4, Figure S5: characteristic of sample 5, Figure S6: schematic figure for the ammonia sensing mechanism of as-prepared sensors, Figure S7: Accurate response of ammonia gas sensor of sample 3, Figure S8: Response and recovery time of sample 3 (Tested under 20 ppm.), Table S1: Difference of samples in the work, Table S2: Comparison of response and recovery time with previous work.

Author Contributions: The experiments and characterizations were carried out by H.W., Z.M., with the assistence of Z.L. and H.S., under the guidance of S.Y. and Y.S., H.W. and S.Y. wrote the manuscript and prepared all figures. Y.S. and S.Y. supervised and coordinated all the work.

Funding: This research was funded by the National Basic Research Program of China (973 Program: 2018YFA0209100), the National Science Foundations of China (No. 61205057, No. 11574136), Qing Lan Project, the "1311 Talent Plan" Foundation of Nanjing University of Posts and Telecommunications, and Six talent peaks project in Jiangsu Province (JY-014).

Conflicts of Interest: The authors declare no competing financial interest.

References

1. Chhowalla, M.; Shin, H.S.; Eda, G.; Li, L.J.; Loh, K.P.; Zhang, H. The chemistry of two-dimensional layered transition metal dichalcogenide nanosheets. *Nat. Chem.* **2013**, *5*, 263–275. [CrossRef] [PubMed]
2. Wang, Y.; Ou, J.Z.; Balendhran, S.; Chrimes, A.F. Electrochemical control of photoluminescence in two-dimensional MoS_2 nanoflakes. *ACS Nano* **2013**, *7*, 10083–10093. [CrossRef] [PubMed]
3. Ning, C.; Qian, L.; Jin, L.; Yong, W.; Bai, Y. Facile synthesis of fluorinated polydopamine/chitosan/reduced graphene oxide composite aerogel for efficient oil/water separation. *Chem. Eng. J.* **2017**, *326*, 17–28.
4. Liu, T.; Yu, K.; Gao, L.; Chen, H.; Wang, N.; Hao, L.; Li, T.; He, H.; Guo, Z. A graphene quantum dot decorated $SrRuO_3$ mesoporous film as an efficient counter electrode for high-performance dye-sensitized solar cells. *J. Mater. Chem. A* **2017**, *5*, 17848–17855. [CrossRef]
5. Choi, W.; Choudhary, N.; Han, G.H.; Park, J.; Akinwande, D.; Lee, Y.H. Recent development of two-dimensional transition metal dichalcogenides and their applications. *Mater. Today* **2017**, *20*, 116–130. [CrossRef]
6. Zhang, F.; Zhu, J.; Zhang, D.; Schwingenschlögl, U.; Alshareef, H.N. Two-dimensional SnO anodes with a tunable number of atomic layers for sodium ion batteries. *Nano Lett.* **2017**, *17*, 1302–1311. [CrossRef] [PubMed]
7. Zhang, D.Z.; Liu, J.J.; Jiang, C.X.; Liu, A.M.; Xia, B.K. Quantitative detection of formaldehyde and ammonia gas via metal oxide-modified graphene-based sensor array combining with neural network model. *Sens. Actuators B Chem.* **2017**, *240*, 55–65. [CrossRef]
8. Li, Y.; Yang, J.; Wang, Y.; Ma, P.; Yuan, Y.; Zhang, J.; Lin, Z.; Zhou, L.; Xin, Q.; Song, A. Complementary integrated circuits based on p-type SnO and n-type IGZO thin-film transistors. *IEEE Electron. Device Lett.* **2018**, *39*, 208–211. [CrossRef]
9. Ogo, Y.; Hiramatsu, H.; Nomura, K.; Yanagi, H.; Kamiya, T.; Hirano, M.; Hosono, H. p-channel thin-film transistor using p-type oxide semiconductor, SnO. *Appl. Phys. Lett.* **2008**, *93*, 032113. [CrossRef]
10. Zhang, Y.; Ma, Z.; Liu, D.; Dou, S.; Ma, J.; Zhang, M.; Guo, Z.; Chen, R.; Wang, S. p-Type SnO thin layers on n-type SnS_2 nanosheets with enriched surface defects and embedded charge transfer for lithium ion batteries. *J. Mater. Chem. A* **2017**, *5*, 512–518. [CrossRef]
11. Hien, V.X.; Lee, J.; Kim, J.; Heo, Y. Structure and NH_3 sensing properties of SnO thin film deposited by RF magnetron sputtering. *Sens. Actuators B Chem.* **2014**, *194*, 134–141. [CrossRef]
12. Kumar, R.; Kushwaha, N.; Mittal, J. Superior, rapid and reversible sensing activity of graphene-SnO hybrid film for low concentration of ammonia at room temperature. *Sens. Actuators B Chem.* **2017**, *244*, 243–251. [CrossRef]
13. Sharma, S.; Kumar, A.; Singh, N.; Kaur, D. Excellent room temperature ammonia gas sensing properties of n-MoS_2/p-CuO heterojunction nanoworms. *Sens. Actuators B Chem.* **2018**, *275*, 499–507. [CrossRef]
14. Timmer, B.; Olthuis, W.; van den Berg, A. Ammonia sensors and their applications—A review. *Sens. Actuators B Chem.* **2005**, *107*, 666–677. [CrossRef]
15. Lamdhade, G.T.; Raulkar, K.B.; Yawale, S.S.; Yawale, S.P. Fabrication of multilayer SnO_2–ZnO–PPy sensor for ammonia gas detection. *Indian J. Phys.* **2015**, *89*, 1025–1030. [CrossRef]
16. Khuspe, G.D.; Navale, S.T.; Bandgar, D.K.; Sakhare, R.D.; Chougule, M.A.; Patil, V.B. SnO_2 nanoparticles-modified Polyaniline Films as Highly Selective, Sensitive, Reproducible and Stable Ammonia Sensors. *Electron. Mater. Lett.* **2014**, *10*, 191–197. [CrossRef]
17. Pathak, A.; Mishra, S.K.; Gupta, B.D. Fiber-optic ammonia sensor using Ag/SnO_2 thin films: optimization of thickness of SnO_2 film using electric field distribution and reaction factor. *Appl. Opt.* **2015**, *54*, 8712–8721. [CrossRef] [PubMed]
18. Yang, F.; Guo, Z. Comparison of the enhanced gas sensing properties of tin dioxide samples doped with different catalytic transition elements. *J. Colloid Interface Sci.* **2015**, *448*, 265–274. [CrossRef] [PubMed]

19. Wang, X.; Li, C.; Huanga, Y.; Zhai, H.; Liu, Z.; Jin, D. Highly sensitive and stable perylene sensor for ammonia detection: A case study of structure-property relationships. *Sens. Actuators B Chem.* **2018**, *275*, 451–458. [CrossRef]
20. Kumar, R.; Kushwaha, N.; Mittal, J. Ammonia gas sensing activity of Sn nanoparticles film. *Sens. Lett.* **2016**, *14*, 300–303. [CrossRef]
21. Du, N.; Zhang, H.; Chen, B.; Ma, X.; Liu, Z.; Wu, J.; Yang, D. Porous Indium Oxide nanotubes: layer-by-layer assembly on Carbon-nanotube templates and application for room-temperature NH3 gas Sensors. *Adv. Mater.* **2007**, *19*, 1641–1645. [CrossRef]
22. Joshi, N.; Hayasaka, T.; Liu, Y.; Liu, H.; Oliveira, O.N., Jr.; Lin, L. A review on chemiresistive room temperature gas sensors based on metal oxide nanostructures, graphene and 2D transition metal dichalcogenides. *Microchim. Acta* **2018**, *185*, 213. [CrossRef] [PubMed]
23. Joshi, N.; da Silva, L.F.; Jadhav, H.S.; Shimizu, F.M.; Suman, P.H.; M'Peko, Je.; Orlandi, M.O.; Seo, J.G.; Mastelaro, V.R.; Oliveira, O.N., Jr. Yolk-shelled $ZnCo_2O_4$ microspheres: Surface properties and gas sensing application. *Sens. Actuators B Chem.* **2018**, *257*, 906–915. [CrossRef]
24. Yang, M.; Chan, H.; Zhao, G.; Bahng, J.H.; Zhang, P.; Král, P.; Kotov1, N.A. Self-assembly of nanoparticles into biomimetic capsid-like nanoshells. *Nat. Chem.* **2017**, *9*, 287–294. [CrossRef] [PubMed]
25. Ghosh, S.; Roy, S. Effect of ageing on $Sn_6O_4(OH)_4$ in aqueous medium—simultaneous production of SnO and SnO_2 nanoparticles at room temperature. *J. Sol–Gel Sci. Technol.* **2017**, *81*, 769–773. [CrossRef]
26. Saji, K.J.; Tian, K.; Snure, M.; Tiwari, A. 2D Tin Monoxide—An unexplored p-Type van der Waals semiconductor: Material characteristics and field effect transistors. *Adv. Electron. Mater.* **2016**, *2*, 1500453. [CrossRef]
27. Kachirayil, J.; Saji, Y.P.; Subbaiah, V.; Tian, K.; Tiwari, A. P-type SnO thin films and SnO/ZnO heterostructures for all-oxide electronic and optoelectronic device applications. *Thin Solid Film* **2016**, *605*, 193–201.
28. Kraya, L.Y.; Liu, G.F.; He, X.; Koel, B.E. Structures and Reactivities of Tin Oxide on Pt(111) Studied by Ambient Pressure X-ray Photoelectron Spectroscopy (APXPS). *Top. Catal.* **2016**, *59*, 497–505. [CrossRef]
29. Liang, L.Y.; Liu, Z.M.; Cao, H.T.; Pan, X.Q. Microstructural, Optical, and Electrical Properties of SnO Thin Films Prepared on Quartz via a Two-Step Method. *ACS Appl. Mater. Interfaces* **2010**, *4*, 1060–1065. [CrossRef] [PubMed]
30. Wu, H.; Zhou, L.; Yan, S.; Song, H.; Shi, Y. Optical Properties of Tin Monoxide Nanoshells Prepared via Self-Assembly. *Nanosci. Nanotechnol. Lett.* **2017**, *9*, 1947–1952. [CrossRef]
31. Malik, R.; Tomer, V.K.; Dankwort, T.; Mishra, Y.K.; Kienle, L. Cubic mesoporous Pd–WO_3 loaded graphitic carbon nitride (g-CN) nanohybrids: highly sensitive and temperature dependent VOC sensors. *J. Mater. Chem. A* **2018**, *6*, 10718–10730. [CrossRef]
32. Paulowicz, I.; Hrkac, V.; Kaps, S.; Cretu, V.; Lupan, O.; Braniste, T.; Duppel, V.; Tiginyanu, I.; Kienle, L.; Adelung, R.; et al. Three-dimensional SnO_2 nanowire networks for multifunctional applications: From high-temperature stretchable ceramics to ultraresponsive sensors. *Adv. Electron. Mater.* **2015**, *1*, 150081. [CrossRef]
33. Postica, V.; Gröttrup, J.; Adelung, R.; Lupan, O.; Mishra, A.K.; de Leeuw, N.H.; Ababii, N.; Carreira, J.F.C.; Rodrigues, J.; Sedrine, N.B.; et al. Multifunctional materials: A case study of the effects of metal doping on ZnO tetrapods with Bismuth and Tin Oxides. *Adv. Funct. Mater.* **2017**, *27*, 1604676. [CrossRef]
34. Mishra, Y.K.; Adelung, R. ZnO tetrapod materials for functional applications. *Mater. Today* **2018**, *21*, 631–635. [CrossRef]

© 2019 by the authors. Licensee MDPI, Basel, Switzerland. This article is an open access article distributed under the terms and conditions of the Creative Commons Attribution (CC BY) license (http://creativecommons.org/licenses/by/4.0/).

Article

Conformational Effects of Pt-Shells on Nanostructures and Corresponding Oxygen Reduction Reaction Activity of Au-Cluster-Decorated NiO$_x$@Pt Nanocatalysts

Dinesh Bhalothia [1,2,†], Yu-Jui Fan [3,†], Yen-Chun Lai [2,4], Ya-Tang Yang [1], Yaw-Wen Yang [4], Chih-Hao Lee [2] and Tsan-Yao Chen [2,5,6,7,*]

1. Institute of Electronics Engineering, National Tsing Hua University, Hsinchu 30013, Taiwan
2. Department of Engineering and System Science, National Tsing Hua University, Hsinchu 30013, Taiwan
3. School of Biomedical Engineering, Taipei Medical University, Taipei 11031, Taiwan
4. National Synchrotron Radiation Research Center, Hsinchu 30007, Taiwan
5. Institute of Nuclear Engineering and Science, National Tsing Hua University, Hsinchu 30013, Taiwan
6. Hierarchical Green-Energy Materials (Hi-GEM) Research Center, National Cheng Kung University, Tainan 70101, Taiwan
7. Higher Education Sprout Project, Competitive Research Team, National Tsing Hua University, Hsinchu 30013, Taiwan
* Correspondence: chencaeser@gmail.com or tsanyao@mx.nthu.edu.tw; Tel.: +886-3-5715131 (ext. 34271); Fax: +885-3-5720724
† These authors contributed equally to this work.

Received: 5 June 2019; Accepted: 9 July 2019; Published: 11 July 2019

Abstract: Herein, ternary metallic nanocatalysts (NCs) consisting of Au clusters decorated with a Pt shell and a Ni oxide core underneath (called NPA) on carbon nanotube (CNT) support were synthesized by combining adsorption, precipitation, and chemical reduction methods. By a retrospective investigation of the physical structure and electrochemical results, we elucidated the effects of Pt/Ni ratios (0.4 and 1.0) and Au contents (2 and 9 wt.%) on the nanostructure and corresponding oxygen reduction reaction (ORR) activity of the NPA NCs. We found that the ORR activity of NPA NCs was mainly dominated by the Pt-shell thickness which regulated the depth and size of the surface decorated with Au clusters. In the optimal case, NPA-1004006 (with a Pt/Ni of 0.4 and Au of ~2 wt.%) showed a kinetic current (J_K) of 75.02 mA cm^{-2} which was nearly 17-times better than that (4.37 mA cm^{-2}) of the commercial Johnson Matthey-Pt/C (20 wt.% Pt) catalyst at 0.85 V vs. the reference hydrogen electrode. Such a high J_K value resulted in substantial improvements in both the specific activity (by ~53-fold) and mass activity (by nearly 10-fold) in the same benchmark target. Those scenarios rationalize that ORR activity can be substantially improved by a syngeneic effect at heterogeneous interfaces among nanometer-sized NiO$_x$, Pt, and Au clusters on the NC surface.

Keywords: oxygen reduction reaction; nanocatalysts; carbon nanotube; wet-chemical reduction method; Au-clusters; mass activity

1. Introduction

Fuel cells are expected to be commercially feasible to moderate deficiencies in natural energy resources without increasing the carbon footprint [1–4]. In spite of many fascinating features like noise-free operation, substantial reductions in pollution, and better efficiencies, the commercial viability of fuel cells is hindered by the substantially high energy barrier of oxygen reduction reactions (ORR) at the cathode [5,6]. To reduce the energy barrier to ORR, platinum (Pt)-based heterogeneous catalysts seem to be the most effective material [7–9]. Due to the unaffordable costs and low storage potential of Pt, finding alternative materials for nanocatalysts (NCs) with comparable efficiencies to Pt is an inevitable step to bringing fuel cells into the market. Meanwhile, lower overpotential losses, long-term durability, pH working conditions, non-toxicity, and earth-abundant elements are fundamental physical and economic requirements for usable material combinations. Despite many efforts so far expended on the development of fuel cells, especially over the past two decades, many hurdles have yet to be overcome. Several Pt-alloys [10–12], including 3D-transition metals (Co, Ni, Cu, Fe, etc.) together with core shell nanostructures [13–15], bimetallic nanodendrites [16], nanowires [17], nano-onions, etc. [18], have been intensively studied in recent decades. In addition, great efforts have been geared towards size [19], shape [20], and composition [21] controls of Pt-based NCs to overcome the aforementioned challenges for preparing highly active ORR catalysts. Those studies laid a strong foundation to further fine-tune the electronic and chemical properties of NCs to extend ORR performances. Promising and efficient techniques, however, are still far away from attaining commercial standards.

Achieving a reconcilable balance between catalytic activity and noble-metal dosages when developing NCs for ORRs is still a challenging task. Core-shell structured heterogeneous NCs with a transition metal (e.g., Co, Ni, Zn, Ru, Fe, and Sn) in the core and a Pt shell seem to be the most effective design in terms of cost considerations and catalytic activities. In such configurations, the core crystal injects electrons (or forms a negative field) to the shell crystal via a combination of three major effects: A bifunctional mechanism [22] (using a variety of adsorption species), ligand effects (electron localization because of the electronegativity gap between two atoms) [23–25], and the lattice strain (differences in atomic arrangements between intraparticle domains) [26–28]. Moreover, such elements, owing to low-energy pathways, provide the opportunity for allocation and recombination kinetics of radicals (i.e., O*, OH*, and H*) in H_2O and reduce durations of intermediate steps on NC surfaces. Meanwhile, the Pt-shell protects transition metals in the core from corrosive conditions at fuel cell cathodes.

The presence of Au in heterogeneous NCs offers electronic, geometric, and compositional effects to tune catalytically active sites that were found to be effective for ORR [29,30]. We further improved the ORR activities of such NCs via decorating strong electronegative atomic-scale Au clusters at the interface and on the surface of Pt-stacked transition metal nanocrystallites. Au clusters not only recover surface defects but also form indirect heterojunctions to the core crystal and localize valence electrons from neighboring atoms using strong electronegative forces. Meanwhile, Pt forms an unconformable shell, which protects the core crystal from corrosion and shares ORR pathways, including O_2 splitting and relocation kinetics of O-atoms, and thus avoids highly energetic intermediates and their associated kinetic penalties. Our previous work demonstrated a facile approach to trigger ORR activity via Pt-decorated core-shell structures [31–35]. Those ternary metallic NCs consisting of lower dosages of Pt showed distinct activity towards ORR facilitation but with reduced fabrication costs.

This study implemented an innovative sequence and time-controlled wet-chemical reduction method to synthesize Au cluster-decorated Ni_{core}-Pt_{shell} NCs. The inner structure (heterogeneous intra-/interparticle interfaces and lattice strain) and surface coverage of such NCs were altered via changing the dosages of Au and Pt. In this event, a series of carbon nanotube (CNT)-supported ternary metallic NCs comprising a Ni/NiO_x core and an Au cluster-decorated Pt-shell (called NPA) were synthesized with variable shell thicknesses (with Pt/Ni ratios of 0.4 and 1.0) and Au contents (~2 and 9 wt.%) on the surface. Such synthesized NCs with unique multiphase cluster-in-cluster interfaces and surface modifications preserved improved catalytic activities towards ORRs in an alkaline environment (0.1 M KOH). Of greatest relevance, the mass activity (MA) and kinetic current density (J_k) of NPA-1004006 (with a Pt/Ni ratio of 0.4 and Au of ~2 wt.%) were improved by 10.36- and 17.16-fold, respectively, compared to those of commercial the Johnson Matthey-Pt/C (20 wt.% Pt) catalyst. At the same time, multiple metallic interfaces contributed to the lower electrochemical surface area (ECSA) with a shift of the Pt oxide reduction peak to more-positive potentials which thus improved the specific activity (SA) by 53.21-fold that of the commercial Johnson Matthey-Pt/C catalyst.

Herein, our findings present a proper strategy for the design of heterogeneous NCs by managing their local structure through controlling the surface and inner configurations. Systematic interpretations of the experimental results are given in latter sections.

2. Experimental

2.1. Synthesis Methodology and Materials for Preparing Ni Core-Pt Shell (Ni@Pt) NCs

Ni@Pt NCs were synthesized using a sequential wet-chemical reduction method [32]. Scheme 1 reveals the reaction steps of the synthesis methodology. Prior to NC synthesis, surface functionalization of catalyst support (multi-walled CNT (MWCNT), Cnano Technology Ltd., Beijing, China) was carried out via acid-treatment in 4.0 M H_2SO_4 at 80 °C for 6 h. In this way, the attachment of metallic crystals onto the CNT surface was strengthened. First, 500 mg of MWCNTs (5 wt.% solution in ethylene glycol (EG)) was added as catalyst support in 1.28 g of an aqueous solution, which contained 0.1 M nickel (II) chloride hexahydrate ($NiCl_2·6H_2O$, Showa Chemical Co. Ltd., Tokyo, Japan), and then this was stirred at 200 rpm for 6 h. The mixture (Ni^{2+} adsorbed CNT; CNT-Ni^{ads}) contained 0.128 mmoles (7.5 mg) of Ni metal ions in a metal loading of Ni/CNT of 30 wt.%. After stirring, 5 mL of a water solution with 0.0386 g $NaBH_4$ (99%, Sigma-Aldrich, St. Louis, MO, USA) was added (step 2) to samples prepared in the first step and stirred at 200 rpm for 10 s. After that, metastable Ni metal nanoparticles (NPs) were formed (Ni/CNT sample). In step 3, 1.28 g of a Pt precursor solution, 0.128 mmoles Pt metal ions (i.e., 0.1 M), was added to the Ni/CNT sample, and then a thin layer of Pt crystals on the surface of nickel NPs (namely Ni@Pt-1010) was formed. In this step, Pt ions could be reduced by an excessive amount of $NaBH_4$ added in step 2 and deposited on the Ni surfaces. According to different Pt/Ni molar ratios, Pt shells with different thicknesses were obtained. The Pt precursor solution was prepared by diluting ~1.0 g $H_2PtCl_6·6H_2O$ (99%, Sigma-Aldrich Co., Burlington, MA, USA) to 18.36 g with distilled water. In the remainder of this article, Ni@Pt NCs with Pt/Ni atomic ratios of 1.0 and 0.4 are called Ni@Pt-1010 and Ni@Pt-1004, respectively.

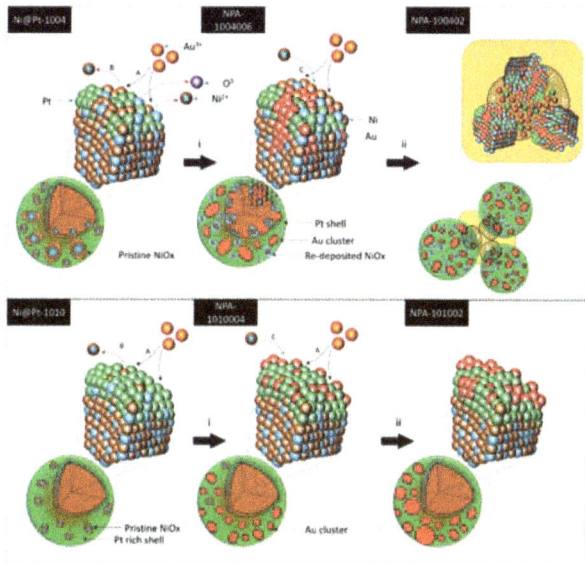

Scheme 1. Crystal growth pathways of Au-decorated Ni@Pt (NPA) nanocatalysts (NCs) with variable Pt/Ni ratios. Among them, pathways A, B, and C respectively correspond to chemisorption of Au^{3+}, galvanic replacement of Au^{3+} with Ni^0, and redeposition of Ni^{2+} on the Ni@Pt surface. Steps i and ii refer to the respective addition of 2.0 and 9.0 wt.% of Au^{3+} to the reaction system.

2.2. Synthesis of Atomic Au Cluster-Decorated Ni Core-Pt Shell (NPA) NCs

A precursor solution of Au was prepared by diluting ~13.3 mg HAuCl$_4$·3H$_2$O (99.0%, Sigma-Aldrich Co., Burlington, MA, USA) to 500 mg with distilled water. After 20 min, once the reaction was complete in step 3, an appropriate amount of the prepared Au solution was added to a solution to make Au/Pt atomic ratios of 0.06 and 0.2. During this stage, Au clusters formed and were intercalated in the Pt shell region through a galvanic replacement reaction of Au^{3+} with Pt, and then with the reducing agent, NaBH$_4$, these metal ions decreased. Finally, products of atomic Au cluster-decorated Ni@Pt (NPA) NCs were obtained. The resulting precipitate was washed several times with acetone, centrifuged, and then dried at 70 °C. In the remainder of this article, Au cluster-deposited Ni@Pt-1004 NCs are called NPA-1004006 and NPA-100402 for 2 and 9 wt.% Au loadings, respectively. Furthermore, Au cluster-deposited Ni@Pt-1010 NCs were named NPA-1010006 and NPA-101002 for 2 and 9 wt.% Au loadings, respectively.

2.3. Physical Characterizations of NPA NCs

Physical properties of the prepared NCs were determined by cross-referencing results of microscopic and X-ray spectroscopic techniques. High-resolution transmission electron microscopic (HRTEM) characterizations were carried out at the electron microscopy center of National Sun Yat-Sen University (Kaohsiung, Taiwan). X-ray photoemission spectroscopy (XPS) of the experimental NCs was executed at beamlines BL-24A1 of the National Synchrotron Radiation Research Center (NSRRC) (Hsinchu, Taiwan). X-ray diffraction (XRD) patterns were collected at an incident X-ray wavelength of 1.5406 Å (8.04 keV) at Taiwan beamline of BL-12B2 in Spring-8 (Hyogo, Japan).

2.4. Preparation of the Electrode and the Method for the ORR Activity Experiment

Catalyst ink for the ORR experiment was made by dispersing 5.0 mg of catalyst powder in a solution consisting of 1.0 mL isopropanol and 50 µL Nafion-117 (99%, Sigma-Aldrich Co., Burlington, MA, USA). This mixture was subjected to ultrasonication for 30 min prior to the ORR test. To conduct the ORR test, 10.0 µL of catalyst ink was drop-cast and air-dried on a glossy carbon rotating disk electrode (RDE) (0.196 cm^2 in area) as the working electrode. The Hg/HgCl$_2$ (with the voltage calibrated to 0.242 V, to align with that of the reference hydrogen electrode (RHE)) electrode saturated in a KCl aqueous solution and a platinum wire were respectively used as the reference and counter electrodes. The ECSAs of the experimental catalysts were calculated by acquiring the columbic charge for reduction of the monolayer Pt oxide after integration and double-layer correction using the following equation:

$$ECSA = \frac{Q_{Pt}}{Q_{ref} \times m}; \quad (1)$$

where Q_{ref} is the charge required for reduction of the monolayer oxide from the bright Pt surface (i.e., 0.405 mC cm^{-2}), m is the metal loading, and Q_{Pt} is the charge required for oxygen desorption, as calculated by following equation:

$$Q_{Pt} = \frac{1}{v}\int (I - I_d)dE. \quad (2)$$

Here, v is the scan rate for the cyclic voltammetric (CV) analysis, and integral parts refer to the area under the Pt oxide reduction peak on CV curves. The kinetic current density (J_K) and number of electrons transferred in ORRs were calculated based on the following equations:

$$\frac{1}{J} = \frac{1}{J_K} + \frac{1}{J_L} = \frac{1}{J_K} + \frac{1}{B\omega^{0.5}} \text{ and} \quad (3)$$

$$B = 0.62nFC_{O_2}D_{O_2}^{\frac{2}{3}}v^{-\frac{1}{6}}; \quad (4)$$

where J, J_K, and J_L are the experimentally measured, mass transport free kinetic, and diffusion-limited current densities, respectively. ω is the angular velocity of the electrode, n is the transferred electron number, F is the Faraday constant, C_{O2} is the bulk concentration of O_2, D_{O2} is the diffusion coefficient, and v is the kinematic viscosity of the electrolyte. For each NC, the MA and SA were respectively obtained when J_K was normalized to the Pt loading and ECSA. Details of the procedure for the ORR mass activity calculation are given in electronic supplementary information (ESI).

2.5. Electrochemical Measurements

Electrochemical measurements were carried out at room temperature (25 ± 1 °C) using a potentiostat (CH Instruments Model 600B, CHI 600B; Hsinchu, Taiwan) equipped with a three-electrode. Cyclic voltammetry (CV) and linear sweep voltammetry (LSV) data were measured at voltage scan rates of 0.02 and 0.001 V s^{-1}, potential ranges of 0.1~1.3 V (vs. the RHE.) and 0.4~1.1 V (vs. the RHE) in an aqueous alkaline electrolyte solution of 0.1 M KOH (pH 13). A rotation rate of 400~3600 rpm was used for LSV. N$_2$ and O$_2$ atmospheres were used for CV and LSV, respectively.

3. Results and Discussion

The particle shape, near-surface configurations, and crystal structure of the experimental NCs were determined using HRTEM analyses. As shown in Figure 1a, Ni@Pt-1004 NPs (Pt/Ni ratio = 0.4) had ordered atomic arrangements at (111) facets exposing the surface (denoted by red arrows). The presence of a clear twin boundary (denoted by a white line) suggested the formation of semi-coherent interfaces. The interplanar spacing of the (111) facet was determined to be 2.239 Å. This value is about 1.25% smaller than that of the Pt-CNT (2.267 Å) indicating the presence of compressive lattice strain in the shell region. With an average particle size of 2.64 nm in (111) facets (determined by an XRD analysis described in a later section), the surface-to-bulk ratio was estimated to be ~50% In this event, considering the ideal case of conformal deposition of Pt atoms by chemisorption and reduction, formation of an incomplete Pt shell over the Ni-core crystal was expected. In contrast, Ni@Pt-1010 NPs (Figure 1d) had grown into isotropic spheres, which comprised a complete Pt-shell and Ni core crystal underneath, and this was also confirmed by XPS analyses in a later section. Compared to that of Ni@Pt-1004 NPs, a higher extent of surface defects (denoted by yellow arrows) was observed in Ni@Pt-1010 NPs. The average particle size of Ni@Pt NCs was observed to be around 2~3 nm, which is consistent with XRD findings in a subsequent section.

Shown in Figure 1b, compared to that of Ni@Pt-1004 NPs, the diameter of NPA-1004006 (i.e., Ni@Pt-1004 NC decorated with 2.0 wt.% of Au atoms) nearly doubled by mixing with Au^{3+} ions for 2 min, followed by the addition of a reducing agent (NaBH$_4$). NPA-1004006 particles grew in a disk-like shape (Figure 1b) with twin boundaries between the NPs. With a nearly identical coherent length to that of Ni@Pt-1004, the large particle was an agglomeration cluster with structural modulations at semi-coherent interfaces between NPs by the galvanic replacement of Au^{3+} to core (Pt/Ni) metals simultaneously with the reduction of residual Pt^{4+}, Ni^{2+}, and Au^{3+} ions by the reducing agents. Such a hypothesis was confirmed by XRD and XPS analyses in later sections. With an incomplete Pt shell structure, certain parts of the NiO$_x$ core were exposed to the liquid environment. In this event, a high content of Ni atoms participated in the galvanic replacement reaction by interacting with Au^{3+} ions. Therefore, Au atoms tended to form metallic clusters between the NPs. As shown in Figure 1c, with a further increase of Au loading to 9 wt.% (NPA-100402), discrete local domains were formed with lattice fringes pointing in different directions (denoted by green arrows). Such a characteristic further revealed severe galvanic replacement of the NiO$_x$ core followed by agglomeration of Ni@Pt NPs by reduction and deposition of residual metal ions in their interfaces. In this event, because of sufficiently high Au loading, the majority of Au atoms were deposited on the top of and between the NPs. Compared to that of Ni@Pt-1004, restructuring of Ni@Pt-1010 by Au^{3+} was insignificant. As shown in Figure 1e,f, particles tended to grow in a core-shell structure with highly ordered atomic arrangements on the surface with increasing Au contents. Such a characteristic was consistently proven by an XRD analysis, which showed that the coherent length of the Pt crystal (111) facet had increased from 2.72 to 3.34 ± 0.1 nm by adding 2.0 wt.% of Au atoms (namely NPA-1010006). Compared to that of NPA-1010006, further increasing Au atoms to 9 wt.% reduced the coherent length to 3.03 ± 0.1 nm, which could be attributed to restructuring between Pt crystals and Au^{3+} ions followed by formation of

Au crystals in the shell region (shown by the presence of a shoulder on the left hand side of the Pt (111) peak). EDX results of NPA-1004006 NC has been shown in Figure S1.

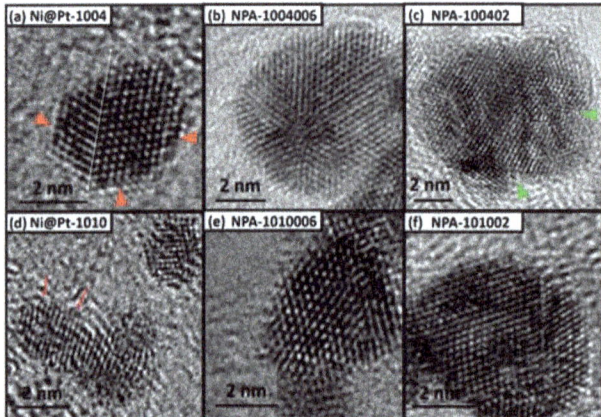

Figure 1. HRTEM images of experimental nanocatalysts (NCs). (**a**) Ni@Pt-1004 NC with Pt/Ni ratios of 0.4 (**b**) and 2.0 wt.% Au-cluster-decorated NI@Pt-1004 NCs (NPA-1004006), (**c**) 9.0 wt.% Au-cluster-decorated NI@Pt-1004 NCs (NPA-100402), (**d**) Ni@Pt-1010 NCs with Pt/Ni ratios of 1.0 (**e**) and 2.0 wt.% Au cluster-decorated Ni@Pt-1010 NCs (NPA-1010006), and (**f**) 9.0 wt.% Au cluster-decorated NI@Pt-1010 NCs (NPA-101002).

Lattice strain, the coherent length (D_{avg}), and the crystallinity of pristine NCs were revealed by XRD analyses. Figure 2 compares XRD patterns of the experimental NCs. Values of the D_{avg} of experimental NCs were calculated from the XRD peak broadening of (111) facets using the Scherrer equation, and their corresponding structural parameters are summarized in Table 1. As indicated in Figure 2, peaks X1 and X2 centered at 40.2° and 46.4° respectively refer to diffraction signals from (111) and (200) facets of metallic-phase Pt nanocrystals in the NCs. In an XRD pattern, the peak width denotes the relative dimension of long-range ordering in specific facets (i.e., the coherent length, D_{avg}) and the ratio of peak intensities between the (111) and (200) facets (i.e., h (111)/h (200)), which refers to the extent of preferential crystal growth for samples under investigation. For Ni@Pt NCs (Figure S2), the D_{avg} increased by 0.1 nm, which could be attributed to the formation of thin layers of Pt in the shell region over the Ni core crystal with Pt/Ni ratios of 0.4 (Ni@Pt-1004) and 1.0 (Ni@Pt-1010) (Table S1). A slight shift in diffraction peaks to the low-angle side features lattice expansion by increasing Pt/Ni ratios from 0.4 to 1.0. Meanwhile, the suppression of NiO_x peaks again consistently revealed increased shell coverage with Pt loading. An even closer inspection of D_{avg} values in (111) and (200) facets reveals the morphology of NCs. The Pt-CNT possessed the highest D (111)/D (200) ratio of 1.33 (higher D (111)) and thus the largest extent of preferential crystal growth along the (111) facet among the experimental NCs. Such a result rationalizes the intrinsic nature of the atomic arrangement in close-packed facets in Pt metal. For Ni@Pt-1004, compared to that of Pt-CNT, the D (111)/D (200) ratio decreased by 0.22, which could be attributed to the competition of crystal growth between galvanic replacement of Pt^{4+} to Ni atoms in the open-(High-Miller-index facets) and Pt deposition in close-packed facets (111). In this status, a semi-coherent lattice match was found between facets with truncated surfaces, which can be explained by the formation of twin boundaries as revealed on HRTEM images (Figure 1a). For Ni@Pt-1010, high contents of Pt^{4+} ion reduction and deposition on the NC surface induced severe galvanic replacement to the NiO_x core. Such a phenomenon mostly occurred at interfaceted corners and edges resulting in a high content of surface truncation in NCs. In this event, NCs tended to form a spherical shape (Figure 1d) due to the strong competition of galvanic replacement to Pt deposition, which was confirmed by a substantially reduced D (111)/D (200) ratio (0.95) compared to that of Pt-CNT.

Table 1. Structural parameters of experimental nanocatalysts (NCs) and control samples as determined by XRD and TEM.

Sample	Pt (111) Facet		Pt (200) facet		Au (111) Facet		Au (200) facet	
	d (Å)	D (nm)	d (Å)	D (nm)	d (Å)	D (nm)	d (Å)	D (nm)
Pt-CNT	2.267	3.52 ± 0.1	1.969	2.63 ± 0.1				
Ni@Pt-1004	2.239	2.64 ± 0.1	1.955	2.38 ± 0.1				
NPA-1004006	2.252	3.03 ± 0.1	1.962	2.50 ± 0.1	2.362	8.24 ± 0.1	2.049	8.09 ± 0.1
NPA-100402	2.266	2.35 ± 0.1	1.965	2.92 ± 0.1	2.352	8.85 ± 0.1	2.043	4.85 ± 0.1
Ni@Pt-1010	2.244	2.72 ± 0.1	1.955	2.86 ± 0.1				
NPA-1010006	2.245	3.34 ± 0.1	1.949	3.01 ± 0.1	2.318	1.75 ± 0.1	2.331	2.63 ± 0.1
NPA-101002	2.254	3.03 ± 0.1	1.960	2.98 ± 0.1	2.046	1.91 ± 0.1	2.046	2.88 ± 0.1

Here "d" and "D" refers to interplanar spacing and average particle size respectively.

After adding different contents of Au to Ni@Pt-1004 NCs (Figure 2a), changes in the lattice strain and crystallinity were obvious due to atomic restructuring. In the case of Ni@Pt-1004 NCs, because of a lower Pt-content (Pt/Ni ratio of 0.4), complete coverage of the Ni-core from Pt atoms was not possible. Thus, with a lesser Au content (NPA-1004006), most of the Au atoms were intercalated with Pt atoms, which aggregated between NCs as revealed by the smeared diffraction peaks across the (111) and (200) facets. Restructuring by formation of Au-rich areas at high-order facets (i.e., (200)) between NCs was significant. These characteristics resulted from the spontaneous trans-metalation between Au^{3+} ions and Ni metal atoms (galvanic replacement) accompanied by redeposition of Ni/Au atoms in shell crystals. Such a phenomenon was consistently proven by the suppressed diffraction signals (peaks M1 and M2) of the Au (111) and (200) facets. Further increasing the Au loading to 9 wt.% (NPA-100402) led to obviously increased diffraction signals for the Au (111) and (200) facets, which revealed the presence of discrete Au clusters (~2 nm) on the surface. At the same time, some of the Au atoms tended to deposit on the shell and core regions. Restructuring of Ni@Pt-1010 by interacting with Au^{3+} ions showed a completely different behavior compared to that of Ni@Pt-1004. As shown in Figure 2b, diffraction peaks X_1 and X_2 were not smeared; on the other hand, they were enhanced by the increasing Au contents from 2.0 to 9.0 wt.%. Such characteristics reveal that the long-range ordering of Ni@Pt NCs was improved by Au decoration. In the absence of diffraction peaks from the Au metal phase and the enhanced Au-4f photoemission peaks (Figure S6, in supplementary information), one can notice that the galvanic replacement of Au^{3+} to the NC surface was suppressed, which suggests conformal deposition of Au atoms in the Pt-rich shell.

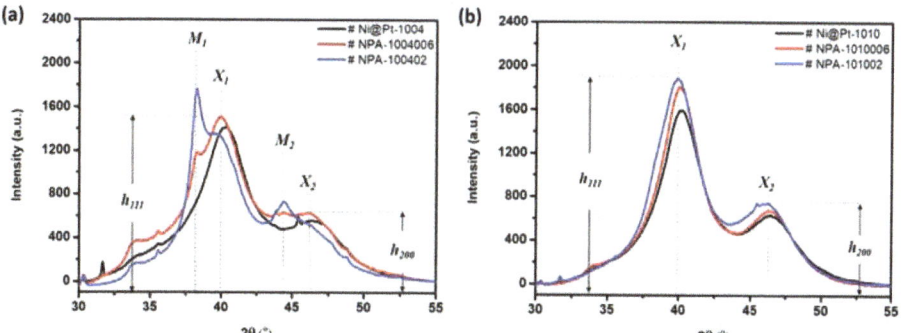

Figure 2. X-ray diffraction patterns of experimental nanocatalysts (NCs) (**a**) NPA-1004006 and NPA-100402 NCs compared to that of the control sample (Ni@Pt-1004), and (**b**) NPA-1010006 and NPA-101002 NCs compared to that of the control sample (Ni@Pt-1010). Peaks "X_1" and "X_2" respectively correspond to diffraction signals from the (111) and (200) facets of NCs (Pt). In contrast, peaks "M_1" and "M_2" respectively correspond to diffraction signals from the (111) and (200) facets of NCs (Au).

The XPS analysis was performed in order to investigate the surface chemical composition (1~2 nm from the surface) and binding energy (BE) of elements in experimental NCs. The incident X-ray with an excitation energy of 650 eV corresponding to a probing depth of ~2.6 nm was employed to probe the Pt-4f, Ni-2p, and Au-4f orbitals. Figure 3 reveals the typical fitted XPS spectra in the Pt-4f region of experimental NCs. In the Pt-4f spectrum, doublet peaks at 71 and 74 eV, respectively, emerged as photoelectron emission lines from the Pt-4f$_{7/2}$ and Pt-4f$_{5/2}$ orbitals. The peaks are further deconvoluted to separate the signals from different oxidation states, and corresponding results are given in Table 2. Through an analysis of XPS patterns of Ni@Pt-1004 and Ni@Pt-1010 NCs (Figure 3a,b), it can be seen that most of the Pt was in a zero-valence state (metallic state). In contrast, from the XPS spectra of Ni-2p orbitals (Figure S3), it is clearly evident that Ni is present in an oxidized (NiO$_x$) form in both Ni@Pt NCs (i.e., Ni@Pt-1004 and Ni@Pt-1010). An even closer inspection of the intensities of the XPS spectra of Ni-2p (Figure S3) revealed that NiO$_x$ signals in Ni@Pt-1010 NCs were very much suppressed compared to those of Ni@Pt-1004 NCs. These spectral characteristics confirmed the core-shell structure of Ni@Pt-1010 NCs comprising a Ni core and Pt in the outermost layer. Meanwhile, the profound intensities of Ni-2p emission peaks in Ni@Pt-1004 NC refer to an incomplete Pt-shell over an underlying Ni-crystal. Those results again confirm prior HRTEM and XRD findings.

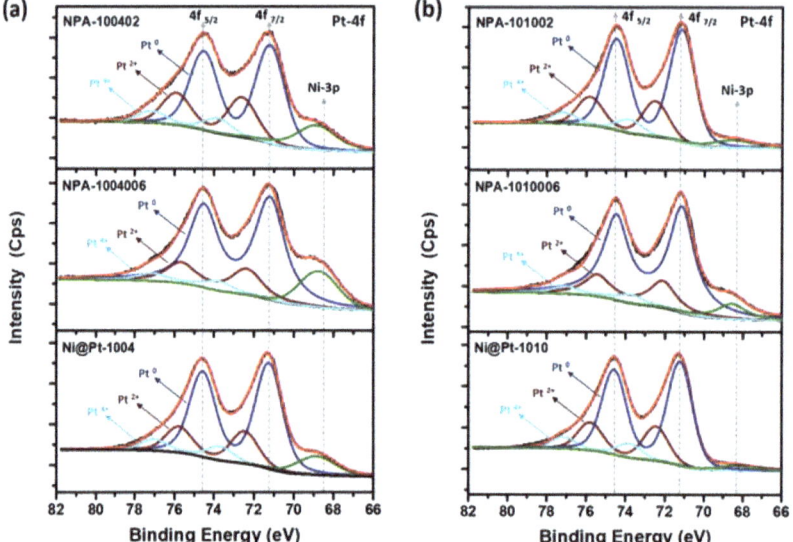

Figure 3. X-ray photoemission spectroscopy of experimental nanocatalysts (NCs). (a) Pt-4f/Ni-3p orbitals of Ni@Pt-1004, NPA-1004006, and NPA-100402. (b) Pt-4f/Ni-3p orbitals of Ni@Pt-1010, NPA-1010006, and NPA-101002.

Table 2. Comparative XPS-determined composition ratios and binding energies of experimental nanocatalysts.

Sample	Binding Energy (eV)			Chemical State Composition (%)		
	Pt^0	Pt^{2+}	Pt^{4+}	Pt^0	Pt^{2+}	Pt^{4+}
Ni@Pt-1004	71.24	72.45	73.68	70	21	9
NPA-1004006	71.20	72.34	73.78	74	19	7
NPA-100402	71.20	72.58	73.8	64	26	10
Ni@Pt-1010	71.23	72.45	73.79	68	22	10
NPA-1010006	71.16	72.11	73.74	75	19	6
NPA-101002	71.15	72.5	73.81	70	22	8

For Au-cluster-decorated Ni@Pt-1004 NCs (Figure 3a), significant restructuring of the surface atomic arrangement was observed. NPA-1004006 NCs (2 wt.% Au) exhibited a higher extent of zero-valent Pt (Pt^0) contents on the surface compared to that of NPA-100402 (9 wt.% Au). More evidence of such atomic restructuring came from the XPS analysis of Ni-2p orbitals (Figure S4). Compared to that of Ni@Pt-1004, we observed that the intensity of the Ni-2p emission peak was gradually suppressed with an increasing amount of Au of 2 to 9 wt.%. Intensities of emission peaks in an XPS spectrum are positively related to the electron density in probing orbitals of target atoms. Therefore, the higher intensity found in the curve of NPA-1004006 revealed that its most abundant 2p electrons were Ni atoms, compared to that of NPA-100402 NCs. Similar spectral changes were observed for Au-cluster-decorated Ni@Pt-1010 NCs (Figure 3b). Obtained XPS results were very consistent with former structural characterizations. The difference in binding energy of the $4f_{7/2}$ orbital of zero-valent Pt (Pt^0) was not obvious in the experimental NCs, revealing that electron relocation between Pt and neighboring atoms was nearly absent. For comparison XPS spectra of Ni@Pt-1010, NPA-1010006 and NPA-101002 at Ni-2p orbitals are compared in Figure S5.

XPS spectra of experimental samples of Au-4f orbitals are compared in Figure S6. Accordingly, intensities of $Au-4f_{5/2}$ and $Au-4f_{7/2}$ peaks increased with Au contents from 2 (NPA-1004006) to 9 wt.% (NPA-100402). Such a phenomenon shows the increasing exposure of Au with loading, which again consistently proves the formation of Au clusters in Ni@Pt NCs with an incomplete shell structure. In the case of Ni@Pt-1010, when the Au loading was 2.0 wt.%, the doublet peaks in the Au-4f orbital were insignificant. Such a result can be complimentarily explained by the crystal structure parameters. As indicated in Figure 2b, a significant improvement in Pt crystallinity was found, and diffraction peak shifts were absent when decorated with 2 wt.% of Au atoms on the Ni@Pt surface. These features indicate the formation of atomic Au clusters on the NiO_x@Pt surface. Therefore, the presence of weak emission peaks suggests a discrete and smeared 4f orbital of Au atomic clusters that were finely dispersed in surface defect sites of NiO_x@Pt NCs. By increasing Au to 9 wt.%, pronounced Au-4f peaks rationalized the formation of Au sub-nano- or nanoclusters, as revealed by presence of a diffraction shoulder on the low-angle side of X1 (111) and a pronounced intensity in X2 (200) peaks.

By cross-referencing results of the physical characterization, the effects of Au^{3+} loading and Pt contents on the evolution of atomic structures of Ni@Pt NCs was systematically determined, and corresponding structural models are given in Scheme 1—where the upper and lower layers respectively present changes in the atomic structure with increasing Au^{3+} loading for Ni@Pt-1004 and Ni@Pt-1010. Accordingly, a significant galvanic replacement on oxidation followed by dissolution of Ni^0 to Ni^{2+} appeared by interacting Ni@Pt-1004 with 2 wt.% of Au^{3+} (step i in the upper layer of Scheme 1) with the reaction of $Au^{3+} + Ni^0$ (or Pt^0) $Au^0 + 3/2Ni^{2+}$ (or $4/3Pt^{4+}$), where Au^{3+} has a higher selectivity for Ni^0 than Pt^0 due to the larger electronegativity difference. In this event, Au atoms tended to penetrate the core region, thus resulting in the coexistence of nanosized Au and Pt clusters in NiO_x@Pt NPs. By increasing the loading to 9 wt.%, galvanic replacement between Au^{3+} ion and core crystals was further enhanced, which caused the severe interparticle agglomeration by dissolution of core metal atoms accompanied by the rapid redeposition of residual metal ions between interfaces (i.e., regions a, b, and c in step ii of the upper layer of Scheme 1) of the NPs. Compared to those of Ni@Pt-1004, the effects of Au^{3+} loading on the structural evolution were suppressed by the high contents of the Pt shell structure in Ni@Pt-1010. As shown in the bottom layer of Scheme 1, Au^{3+} tended to be adsorbed and was reduced by $NaBH_4$ to form atomic clusters on the NP surface with a loading of 2.0 wt.% (NPA-1010006). By increasing the loading to 9.0 wt.%, the Au^{3+} ions tended to form homoatomic bonds and thus grew into sub-nanometer crystals on the NPA-101002 surface (step ii in the bottom layer of Scheme 1). These atomic structural arrangements provide direct information explaining ORR activities of the experimental NCs.

In heterogeneous catalysts, dissociation of chemisorbed oxygen molecules (i.e., the oxygen adsorption strength) is a cardinal performance-determining factor in ORRs. Lowering the oxygen adsorption energy reduces the applied energy for initiating ORRs at reaction sites and relocating them to neighboring atoms. In this way, the reaction kinetics and ORR activities of NCs can be substantially improved. In this study, the surface composition design of catalysts within the sub-nano scale played a key role in ORR performances of the NCs. By cross-referencing physical inspection results (the upper layer of Scheme 1), the surface chemical configuration of NiO$_x$@Pt comprised mixtures of sub-nano Au and Pt clusters in the shell region and Au clusters intercalated with NiO$_x$ in the core when the Pt/Ni ratio was 0.4 and Au was 2.0 wt.% With an Au content of 9.0 wt.%, severe interparticle agglomeration due to galvanic replacement (Au^{3+} + Pt/Ni → Au + Pt^{4+}/Ni^{2+}) accompanied by Au crystal growth and redeposition of residual metal ions between NCs occurred. All three pathways dramatically reduced the degree of heteroatomic intermixing on the surface among reaction sites; therefore, ORR activities of those NCs were substantially suppressed.

Results of the electrochemical analyses consistently elucidated the above scenarios. Figure 4a compares CV curves of the commercial Johnson Matthey-Pt/C catalyst (Johnson Matthey-Pt/C) with the experimental NCs (i.e., Ni@Pt-1004, NPA-1004006, and NPA-100402). Electrochemical active surface areas (ECSAs) are calculated based on corresponding CV curves using the oxygen desorption peak in the backward potential sweeping curve (detailed ECSA data of various ORR catalysts are listed in electronic supplementary information in Table S2). Three distinctive potential regions are found in a CV curve, including an under-potential deposition of hydrogen (UPD-H) region at 0 < E < 0.4 V, a double-layer region between 0.4 and 0.6 V, and chemisorption of oxygen species at >0.6 V vs. the RHE because of hydrogen adsorption/desorption, OH- ligand chemisorption, and the formation of alpha Pt oxide (E_O^{ads}; forward scan) as well as a reduction in Pt oxides (E_O^{des}; backward scan). In this way, the position and width of each peak are susceptible to the chemical composition and structure of the NC surface. For the Johnson Matthey-Pt/C, positions of two characteristic peaks (H_1 and H_2) in the forward scan denoted the potential to be applied for dissociation of H$^+$ from close-packed (111) and opened (200) facets and the corresponding current, respectively. In contrast, peaks H_1^* and H_2^* respectively refer to current responses of H$^+$ adsorption in the (111) and (200) facets. For Ni@Pt-1004, compared to the CV profile of the Johnson Matthey-Pt/C, a downshift of peaks H_1 and H_2 in the forward scan and an upshift of peaks H_1^* and H_2^* in the backward scan refer to a decreased energy barrier for redox desorption/adsorption of H$^+$. As consistently shown by XRD observations, a substantially higher intensity of peak H_1 than that of peak H_2 (i.e., weakened H$^+$ interactions on (200) facets) revealed preferential crystal growth at (111) facets in all experimental NCs.

Compared to that of the Johnson Matthey-Pt/C, Ni@Pt-1004 showed a higher surface area for H$_2$ evolution as revealed by the larger area of the UPD_H region. Meanwhile, the broadened and smeared UPD_H peaks revealed a high density of surface defects in Ni@Pt-1004. This statement is consistently illustrated by the pronounced oxygen adsorption peak (E_O^{ads}) with a potential shift by ca. ~0.17 V (i.e., easy oxidation of the Ni@Pt-1004 surface) compared to that of the Johnson Matthey-Pt/C. Compared to the CV profile of NCs without Au decoration, a slight amount of Au decoration reduced the surface defect sites of Ni@Pt-1004, as consistently revealed by the significant suppression of the E_O^{ads} and E_O^{des} peaks in NPA-1004006. Moreover, the position of the oxide reduction peak (E_O^{des}) was upshifted to high-potential sites at the same time. These two observations integrally bring out the fact that atomic decoration by Au clusters can fix the defect and simultaneously suppress the surface oxidation on the Ni@Pt-1004 surface. In the presence of discrete 4f orbitals in atomic clusters, a strong repulsive force to the chemisorbed O (O^{ads}) was formed at Au atoms. This relocated the O^{ads} to neighboring atoms around the Pt and Au interfaces and thus substantially boosted the ORR kinetic current (J_k) of NPA-1004006 to ~75 mA cm^{-2} (details discussed below in the LSV analysis, Figure 4c). For NPA-100402, the H$^+$ adsorption peak, "H_1^*", in the backward scan together with a smeared CV profile in forward scan was observed in the UPD_H region. Such a feature can be rationalized by its nanostructure, where the surface of the NC consists of nanosized Au/Pt clusters. Meanwhile, a severe interparticle agglomeration by the strong galvanic replacement accompanied by rapid reduction of residual metal ions was found (Figure 1c, Scheme 1); therefore, the heteroatomic intermix and the amount of reaction sites dramatically decreased. Formation of cluster-in-cluster structures turned 4f orbitals from a discrete state into a band structure. In this state, both the Au and Pt atoms possessed bulk properties, and consequently the J_k of NPA-100402 was dramatically reduced by 87% (65.62 mA cm^{-2}) to 9.4 mA cm^{-2} (Table S2), compared to that of NPA-1004006. This value was even lower than that of Ni@Pt-1004, indicating that the impact of the heterogeneous interface between nanoclusters on ORR activity was limited.

On the other hand, as shown in Figure 4b, changes in profiles in CV curves were insignificant by decoration with Au atoms on Ni@Pt-1010. In the case of 2.0 wt.% Au decoration, a slight suppression of both E_O^{ads} and E_O^{des} was found for NPA-1010006 NCs compared to that of Ni@Pt-1010. These characteristics resemble the same redox response to O adsorption and desorption when decorating Au atomic clusters on the Ni@Pt-1004 surface. Increasing the Au content to 9.0 wt.% did not further suppress the redox peaks of O evolutions (E_O^{ads}/E_O^{des}); however, they moved in an opposite direction. Due to the strong preference for homoatomic bonding, Au atoms tended to form sub-nano clusters instead of a conformal coating or defect sites on NC Pt surfaces. Those characteristics reduced the heteroatomic intermix and amount of O_2-splitting sites in NCs. The former suppresses intrinsic activity, and the latter reduces the number of active sites; therefore, the J_k of NPA-101002 was substantially reduced by 70% (to ~22 mA cm^{-2}), compared to that of NPA1010006. Again, given that most of the Au atoms were deposited as sub-nanometer clusters or atomic clusters (shown by absence of an Au diffraction peak in the XRD pattern of Figure 2b), a high J_k (~22 mA cm^{-2}) was expected.

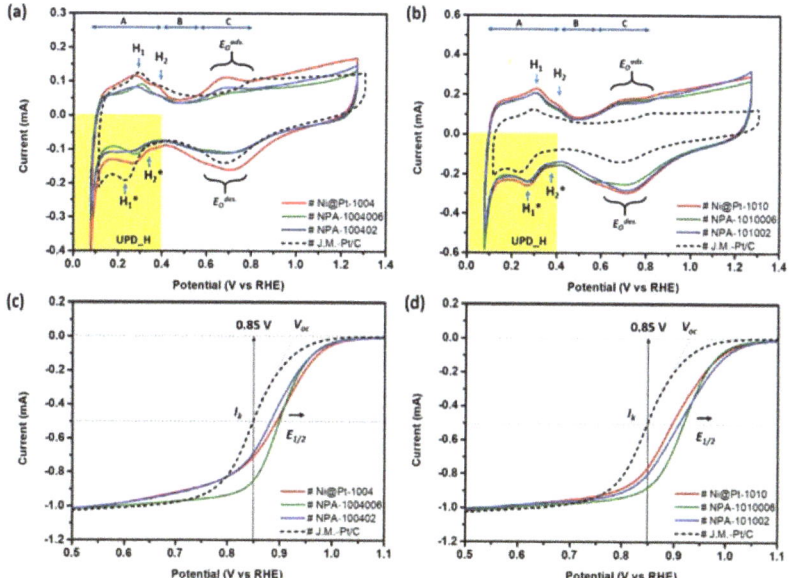

Figure 4. Electrochemical analysis of experimental nanocatalysts (NCs). (a,b) CV and (c,d) LSV curves of experimental NCs compared to a commercial Johnson Matthey-Pt/C NC. Potential sweeping curves were measured at a rotation speed of the electrode of 1600 rpm.

To further rationalize the impacts of Au decoration and Pt contents on ORR activities of Ni@Pt NPs, an LSV analysis was employed. Figure 4c,d demonstrates LSV curves of NPA compared to control samples (Ni@Pt-1004 and Ni@Pt-1010) and the Johnson Matthey-Pt/C, where corresponding electrochemical parameters are summarized in Figure 5, and Tables S2 and S3. As shown in Figure 4c and Table S2, the onset potential (Voc vs. the RHE) of experimental samples followed the trend of the Johnson Matthey-Pt/C (0.910 V) < NPA-1004006 (0.964 V) < NPA-100402 (0.969 V) < Ni@Pt-1004 (0.990 V). Among them, the highest Voc indicates the lowest activation energy for initiating ORRs on Ni@Pt-1004 surfaces and can be rationalized by a large extent of heteroatomic intermix between Pt and Ni sub-nanoclusters. On such a surface, Pt clusters are in charge of splitting (dissociating) the oxygen molecule (O_2). After O_2 splitting, the two chemisorbed oxygen atoms are relocated to the high-oxygen-affinity Ni atoms for the subsequent reduction reaction. Since the dimensions of Pt clusters are quite small (i.e., the interface ratio between Ni domain is high), a rapid relocation of chemisorbed oxygen atoms to Ni sites is expected; therefore, the J_k of Ni@Pt-1004 doubled compared to that of the Johnson Matthey-Pt/C. By decorating with 2.0 wt.% Au atoms on Ni@Pt-1004, the atomic Au clusters reduced surface defects of the Pt shell and covered the Ni core crystal. In this event, activation energy for O_2 splitting increased, which reduced the intrinsic activities of surface sites and thus the Voc by 0.026 V. Such a phenomenon would seemingly suppress the activity of surface sites on the NC surface; however, it actually went the opposite way. As revealed by results of physical structural inspections, decoration with a slight amount of Au atoms (2.0 wt.%) resulted in a surface structure comprising combinations of Au, Pt, and Ni sub-nanoclusters on the NC surface. The presence of sub-nano or atomic Au clusters in defect sites protected the NC surface from oxidation (shown by the suppressed E_O^{ads} peak in the CV curve). In this event, relocation of the O^{ads} was dramatically facilitated, resulting in a quantum leap of J_k by 6.7-fold compared to that of NCs without Au decoration (i.e., Ni@Pt-1004). By increasing Au to 9.0 wt.%, compared to that of NPA-1004006, the J_k of NPA-100402 substantially decreased by ~87% to 9.4 mAcm^{-2}. This value was almost the same as that of Ni@Pt-1004, indicating that all reaction sites (Pt, Au, and Ni) possessed metallic properties of NPs (bulk) without facilitation of heteroatomic intermixing or ligand effects in ORRs. Meanwhile, as shown in Figure 4c, similar slopes of diffusion (V < 0.8 V vs. the RHE) and kinetic limit (V > 0.8 V vs. the RHE) current regions indicated that the redox responses of Ni@Pt-1004 were not greatly influenced by a high Au loading. Such electrochemical properties are understandable due to the fact that Au atoms tend to form large nanocrystals by homogeneous crystal growth on the surface and galvanic replacement in the core region of Ni@Pt NCs. Compared to those of Ni@Pt-1004, Au decoration showed similar effects on ORR activity of Ni@Pt NCs with a conformal Pt shell (Ni@Pt-1010). Figure 4d compares LSV curves of experimental NCs (NPA-1010006 and NPA-101002) with those of the control sample (Ni@Pt-1010) and the Johnson Matthey-Pt/C; corresponding electrochemical properties are summarized in Table S3. Accordingly, VOC and $E_{1/2}$ followed the same trend as that of Ni@Pt-1004 with increasing Au contents from 2.0 to 9.0 wt.%. The same scenario to NPA NCs with low Pt contents held for NPA-1010006, except that Au atoms did not penetrate into the core region to form nanosized clusters, therefore NPA-101002 retained a high J_k value (21.99 mA cm^{-2}) in ORRs.

Surface activities (SAs) of electrocatalysts in ORRs are calculated by normalizing the J_k to the ECSA of the oxygen desorption region in the CV curve. These values are an important index for the average intrinsic activity of reaction sites on NC surfaces. As shown in Table S2, the ECSA was 81.2 cm^2 mg^{-1} for Ni@Pt-1004 and decreased to 50.0 cm^2 mg^{-1} by decorating with 2.0 wt.% Au atoms, again proving that Au atoms tended to reside in defect sites of the Pt shell. Further increasing the Au content to 9.0 wt.% did not affect the ECSA value. This result, consistent with that proven by the XRD analysis, suggests that Au atoms tended to grow in homoatomic nanocrystals instead of capping on the Pt shell surface. Accordingly, SAs were 13.89 mA cm^{-2} for NPA-1004006 and 1.83 mA cm^{-2} for NPA-100402. With similar ESCAs, the significantly enhanced SA elucidates conformation of the substantially improved intrinsic ORR activity by the presence of atomic/sub-nano Au clusters simultaneously with Pt and Ni nanoclusters on the NPA-1004006 surface. For Ni@Pt with a conformal Pt shell, Au decoration mainly occurred on the NPA-1010006 surface, as indicated by a reduction in the ECSA (64.4 cm^2 mg^{-1}) by 11.9% compared to that of Ni@Pt-1010 (73.1 cm^2 mg^{-1}). Compared to that of NPA-1010006, the ESCA increased by 11.3% when decorated with 9.0 wt.% of Au on the Ni@Pt-1010 surface, which can be attributed to the formation of sub-nano Au clusters on the NC surface. In this event, SAs were 5.64 mA cm^{-2} for NPA-1010006 and 1.61 mA cm^{-2} for NPA-101002. Compared to those of NPA with Pt/Ni = 0.4, the substantially suppressed SA reveals the truth that the combination of atomic Au/Pt clusters with Ni atoms in neighboring sites can support exceptional reaction kinetics of ORRs (i.e., J_k and SA).

Mass activity (MA) refers to the current density per unit weight of active sites and is calculated by normalizing the residual current at 0.85 V vs. the RHE with respect to the loading amount of metal Pt in NCs. As illustrated in Figure 5a, the MA of Ni@Pt was slightly improved by 26% compared that of the Johnson Matthey-Pt/C. By adding 2.0 wt.% Au atoms on the surface, the MA of NPA-1004006 substantially improved by 7.1-fold compared to that of Ni@Pt-1004. With a slight increment of noble metal loading, such a dramatic improvement in the MA depicts the truth for boosting the activity of NCs by syngeneic collaboration between sub-nano Au, Pt, and Ni domains in the reaction pathways in ORRs. Such a scenario was further confirmed by the MA of NPA-100402. In this NC, the MA was significantly reduced by 87% compared to that of NPA-1004006. Given that the difference in noble metal loading was small (7.0 wt.%), the dramatic difference in the MA again proves changes in intrinsic activities instead of mass differences. The same phenomenon exists in changes of the MA with respect to Au loading in Ni@Pt-1010 again complementarily proves the synergetic effects on ORR activities of NCs. Electrochemical results of control samples with commercial catalyst has been compared in Figure S7 and corresponding parameters has been summarized in Table S4.

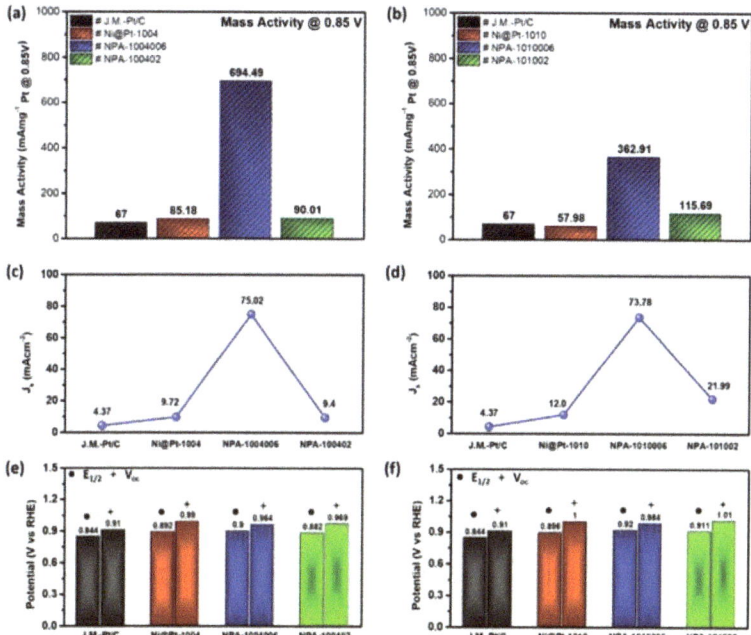

Figure 5. Electrochemical results of experimental nanocatalysts (NCs). (a,b) Oxygen reduction reaction (ORR) mass activity of NCs compared to the commercial Johnson Matthey-Pt/C at 0.85 V (vs. the reference hydrogen electrode (RHE)). (c,d) Kinetic current density of the experimental NCs and Johnson Matthey-Pt/C at 0.85 V (vs. the RHE). (e,f) Onset potential and half-wave potential of the experimental NCs and Johnson Matthey-Pt/C.

4. Conclusions

CNT-supported NCs with a Ni/NiO$_x$ base and an Au cluster-modified Pt-shell were synthesized via self-aligned wet-chemical processes with variable shell thicknesses (Pt/Ni ratios of 0.4 and 1.0) decorated with different contents of Au atoms (2 and 9 wt.%). Results of physical structural characterizations combined with electrochemical analyses proved that surface coverage of the Pt-shell along with depth and distribution of Au clusters significantly affected inner structural configurations and thus the ORR activities of bimetallic Ni@Pt NCs. For Ni@Pt-1004 NCs, because a lower Pt-content and lower surface coverage were adopted, Au atoms tended to form sub-nanoclusters accompanied by Pt and Ni on the NC surface at a loading of 2.0 wt.% Such an NC exhibited the highest J_k (75.02 mA cm^{-2}), SA (13.89 mA cm^{-2}), and MA (694.49 mA mg$_{Pt}^{-1}$) among the experimental NCs due to the synergetic collaboration between oxygen-inert and -affinity sites on the surface. When increasing the loading to 9.0 wt.%, Au atoms tended to penetrate into the core region and grow into homoatomic clusters on the NC surface. Both characteristics reduced the heteroatomic intermix and surface ratio, and therefore, turned the redox properties of NCs into a bulk nanocrystal state. For the case of Ni@Pt with a conformal shell, Au atoms tended to form atomic clusters on the NC surface which exhibited a J_k of 73.78 mA cm^{-2} corresponding to an MA of 362.9 mA mg$_{Pt}^{-1}$ and SA of 5.64 mA cm^{-2}. Compared to those with low Pt contents, the lower electrochemical performances of NPA with a Pt/Ni ratio of 1.0 consistently explained the local syngeneic effects on the NC surface. In brief, robust methods to synthesize bimetallic NCs with different extents of surface decoration were developed in this study. We demonstrated that such processes can be adopted to control the identity and local structural disorder on NC surfaces. By proper control of the surface decoration loading, the ORR performance of NiO@Pt catalysts was improved by 7.1-fold in the optimal case. These results elucidate a new prospect of heterogeneous catalyst design. It realizes a compact co-catalyst with different sub-nanometer components collaborating together to share intermediate steps in redox reactions and thus enabling facilitation of the activity of electrocatalysts.

Supplementary Materials: The following are available online at http://www.mdpi.com/2079-4991/9/7/1003/s1. Figure S1: TEM images of NPA-1004006 NC. The EDX results are measured in the region marked by pink square. Figure S2: XRD patterns of Ni@Pt-1004 and Ni@Pt-1010 experimental NCs. Figure S3: Comparative x-ray photoemission spectroscopy of Ni@Pt-1004 and Ni@Pt-1010 NCs at Ni-2p orbitals. Figure S4: X-ray photoemission spectroscopy of Ni@Pt-1004, NPA-1004006 and NPA-100402 NCs at Ni-2p orbitals. Figure S5: X-ray photoemission spectroscopy of Ni@Pt-1010, NPA-1010006 and NPA-101002 NCs at Ni-2p orbitals. Figure S6: X-ray photoemission spectroscopy of experimental NCs. (a) Au-4f orbitals of NPA-1004006 and NPA-100402. (b) Au-4f orbitals of NPA-1010006 and NPA-101002. Figure S7: Electrochemical analysis of experimental NCs. (a) CV and (2) LSV curves of Ni@Pt-1004 and Ni@Pt-1010 experimental NCs compared with commercial J.M.-Pt/C nanocatalysts. Table S1: XRD and TEM determined structural parameters of control samples. Table S2: Electrochemical parameters of experimental NCs with Pt/Ni = 0.4. Table S3: Electrochemical parameters of experimental NCs with Pt/Ni = 1.0. Table S4: Electrochemical parameters of control samples.

Author Contributions: T.Y.C. conceived the experiments and designed the synthetic procedures of the ternary nanocatalyst. Y.C.L., Y.J.F., C.H.L., and Y.W.Y. Prepared various catalyst samples and performed the electrochemical measurements. All authors analyzed the experimental data. D.B, Y.T.Y, and T.Y.C. wrote the manuscript and prepared the figures. All authors participated in discussions and knew implications of the work.

Funding: This research received no external funding.

Acknowledgments: The authors thank the staff of the National Synchrotron Radiation Research Center (NSRRC), Hsinchu, Taiwan and Spring-8, Japan for helping with various synchrotron-based measurements and XRD analyses. T.-Y. Chen acknowledges the funding support from National Tsing Hua University, Taiwan (N103K30211 and 103N1200K3) and the Ministry of Science and Technology, Taiwan (MOST 106-2112-M-007-016-MY3 and MOST 105-3113-E-006-019-CC2).

Conflicts of Interest: The authors declare no conflict of interest.

References

1. Das, V.; Padmanaban, S.; Venkitusamy, K.; Selvamuthukumaran, R.; Blaabjerg, F.; Siano, P. Recent advances and challenges of fuel cell based power system architectures and control—A review. *Renew. Sustain. Energy Rev.* **2017**, *73*, 10–18. [CrossRef]
2. Hansen, J.; Sato, M.; Ruedy, R.; Kharecha, P.; Lacis, A.; Miller, R.; Nazarenko, L.; Lo, K.; Schmidt, G.A.; Russell, G.; et al. Dangerous human-made interference with climate: A GISS modelE study. *Atmos. Chem. Phys.* **2007**, *7*, 2287–2312. [CrossRef]
3. Holton, O.T.; Stevenson, J.W. The Role of Platinum in Proton Exchange Membrane Fuel Cells. *Platin. Met. Rev.* **2013**, *57*, 259–271. [CrossRef]

4. Sasaki, K.; Naohara, H.; Cai, Y.; Choi, Y.M.; Liu, P.; Vukmirovic, M.B.; Wang, J.X.; Adzic, R.R. Core-Protected Platinum Monolayer Shell High-Stability Electrocatalysts for Fuel-Cell Cathodes. *Angew. Chem. Int. Ed.* **2010**, *49*, 8602–8607. [CrossRef] [PubMed]
5. Larminie, J.; Dicks, A. *Fuel Cell Systems Explained*, 2nd ed.; John Wiley & Sons Ltd.: Chichester, UK, 2003; pp. 45–66.
6. Vielstich, W.; Lamm, A.; Gasteiger, H.A. *Handbook of Fuel Cells: Fundamentals, Technology, and Applications*; John Wiley & Sons: Hoboken, NJ, USA, 2009; Volume 5.
7. Hernandez-Fernandez, P.; Masini, F.; McCarthy, D.N.; Strebel, C.E.; Friebel, D.; Malacrida, P.; Nierhoff, A.; Bodin, A.; Wise, A.M.; Nielsen, J.H.; et al. Mass-selected of PtxY as model catalysts for oxygen electroreduction. *Nat. Chem.* **2014**, *6*, 732–738. [CrossRef]
8. Mench, M.M.; Kumbar, E.C.; Veziroglu, T.N. *Polymer Electrolyte Fuel Cell Degradation*; Academic Press Elsevier: Waltham, MA, USA, 2012.
9. Zhang, S.; Yuan, X.; Hin, J.N.C.; Wang, H.; Friedrich, K.A.; Schulze, M. A review of platinum-based catalyst layer degradation in proton exchange membrane fuel cells. *J. Power Sources* **2009**, *194*, 588–600. [CrossRef]
10. Huang, X.; Zhao, Z.; Cao, L.; Chen, Y.; Zhu, E.; Lin, Z.; Li, M.; Yan, A.; Zettl, A.; Wang, Y.M.; et al. High-performance transition metal–doped Pt3Ni octahedra for oxygen reduction reaction. *Science* **2015**, *348*, 1230. [CrossRef] [PubMed]
11. Stamenkovic, V.R.; Fowler, B.; Mun, B.S.; Wang, G.; Ross, P.N.; Lucas, C.A.; Marković, N.M. Improved Oxygen Reduction Activity on Pt3Ni(111)via Increased Surface Site Availability. *Science* **2007**, *315*, 493. [CrossRef]
12. Dong, Y.; Zhou, Y.-W.; Wang, M.-Z.; Zheng, S.-L.; Jiang, K.; Cai, W.-B. Facile Aqueous Phase Synthesis of Carbon Supported B-doped Pt3Ni Nanocatalyst for Efficient Oxygen Reduction Reaction. *Electrochim. Acta* **2017**, *246*, 242–250. [CrossRef]
13. Jiang, J.; Gao, H.; Lu, S.; Zhang, X.; Wang, C.-Y.; Wang, W.-K.; Yu, H.-Q. Ni–Pd core–shell nanoparticles with Pt-like oxygen reduction electrocatalytic performance in both acidic and alkaline electrolytes. *J. Mater. Chem. A* **2017**, *5*, 9233–9240. [CrossRef]
14. Luo, L.; Zhu, F.; Tian, R.; Li, L.; Shen, S.; Yan, X.; Zhang, J. Composition-Graded Pd_xNi_{1-x} Nanospheres with Pt Monolayer Shells as High-Performance Electrocatalysts for Oxygen Reduction Reaction. *ACS Catal.* **2017**, *7*, 5420–5430. [CrossRef]
15. Strasser, P.; Kühl, S. Dealloyed Pt-based core-shell oxygen reduction electrocatalysts. *Nano Energy* **2016**, *29*, 166–177. [CrossRef]
16. Lim, B.; Jiang, M.; Camargo, P.H.C.; Cho, E.C.; Tao, J.; Lu, X.; Zhu, Y.; Xia, Y. Pd-Pt Bimetallic Nanodendrites with High Activity for Oxygen Reduction. *Science* **2009**, *324*, 1302. [CrossRef] [PubMed]
17. Li, M.; Zhao, Z.; Cheng, T.; Fortunelli, A.; Chen, C.Y.; Yu, R.; Zhang, Q.; Gu, L.; Merinov, B.V.; Lin, Z.; et al. Ultrafine Jagged Platinum Nanowires Enable Ultrahigh Mass Activity for the Oxygen Reduction Reaction. *Science* **2016**, *354*, 1414–1419. [CrossRef] [PubMed]
18. Choi, E.Y.; Kim, C.K. Fabrication of nitrogendoped nano-onions and their electrocatalytic activity toward the oxygen reduction reaction. *Sci. Rep.* **2017**, *7*, 4178. [CrossRef] [PubMed]
19. Nesselberger, M.; Ashton, S.; Meier, J.C.; Katsounaros, I.; Mayrhofer, K.J.J.; Arenz, M. The Particle Size Effect on the Oxygen Reduction Reaction Activity of Pt Catalysts: Influence of Electrolyte and Relation to Single Crystal Models. *J. Am. Chem. Soc.* **2011**, *133*, 17428–17433. [CrossRef] [PubMed]
20. Shao, M.; Chang, Q.; Dodelet, J.-P.; Chenitz, R. Recent Advances in Electrocatalysts for Oxygen Reduction Reaction. *Chem. Rev.* **2016**, *116*, 3594–3657. [CrossRef]
21. Wang, Y.-J.; Zhao, N.; Fang, B.; Li, H.; Bi, X.T.; Wang, H. Carbon-Supported Pt-Based Alloy Electrocatalysts for the Oxygen Reduction Reaction in Polymer Electrolyte Membrane Fuel Cells: Particle Size, Shape, and Composition Manipulation and Their Impact to Activity. *Chem. Rev.* **2015**, *115*, 3433–3467. [CrossRef]
22. Park, S.A.; Lee, E.K.; Song, H.; Kim, Y.T. Bifunctional enhancement of oxygen reduction reaction activity on Ag catalysts due to water activation on LaMnO3 supports in alkaline media. *Sci. Rep.* **2015**, *5*, 13552. [CrossRef]
23. Bligaard, T.; Nørskov, J.K. Ligand effects in heterogeneous catalysis and electrochemistry. *Electrochim. Acta* **2007**, *52*, 5512–5516. [CrossRef]
24. Gauthier, Y.; Baudoing, R.; Joly, Y.; Rundgren, J.; Bertolini, J.C.; Massardier, J. Pt_xNi_{1-x}(111) alloy surfaces: Structure and composition in relation to some catalytic properties. *Surf. Sci.* **1985**, *162*, 342–347. [CrossRef]

25. Greeley, J.; Nørskov, J.K.; Mavrikakis, M. Electronic structure and catalysis on metal surfaces. *Annu. Rev. Phys. Chem.* **2002**, *53*, 319–348. [CrossRef] [PubMed]
26. Stamenkovic, V.; Mun, B.S.; Mayrhofer, K.J.J.; Ross, P.N.; Markovic, N.M.; Rossmeisl, J.; Greeley, J.; Nørskov, J.K. Changing the Activity of Electrocatalysts for Oxygen Reduction by Tuning the Surface Electronic Structure. *Angew. Chem. Int. Ed.* **2006**, *45*, 2897–2901. [CrossRef] [PubMed]
27. Paffett, M.T.; Daube, K.A.; Gottesfeld, S.; Campbell, C.T. Electrochemical and surface science investigations of PtCr alloy electrodes. *J. Electroanal. Chem. Interfac. Electrochem.* **1987**, *220*, 269–285. [CrossRef]
28. Hammer, B.; Nørskov, J.K. Electronic factors determining the reactivity of metal surfaces. *Surf. Sci.* **1995**, *343*, 211–220. [CrossRef]
29. Kang, Y.; Snyder, J.; Chi, M.; Li, D.; More, K.L.; Markovic, N.M.; Stamenkovic, V.R. Multimetallic Core/Interlayer/Shell Nanostructures as Advanced Electrocatalysts. *Nano Lett.* **2014**, *14*, 6361–6367. [CrossRef]
30. Shen, L.-L.; Zhang, G.-R.; Miao, S.; Liu, J.; Xu, B.-Q. Core–Shell Nanostructured Au@NimPt2 Electrocatalysts with Enhanced Activity and Durability for Oxygen Reduction Reaction. *ACS Catal.* **2016**, *6*, 1680–1690. [CrossRef]
31. Chen, H.Y.T.; Chou, J.P.; Lin, C.Y.; Hu, C.W.; Yang, Y.T.; Chen, T.Y. Heterogeneous Cu-Pd Binary Interface Boosts Stability and Mass Activity of Atomic Pt Clusters in the Oxygen Reduction Reaction. *Nanoscale* **2017**, *9*, 7207–7216. [CrossRef]
32. Zhuang, Y.; Chou, J.-P.; Tiffany Chen, H.-Y.; Hsu, Y.-Y.; Hu, C.-W.; Hu, A.; Chen, T.-Y. Atomic scale Pt decoration promises oxygen reduction properties of Co@Pd nanocatalysts in alkaline electrolytes for 310k redox cycles. *Sustain. Energy Fuels* **2018**, *2*, 946–957. [CrossRef]
33. Bhalothia, D.; Chou, J.-P.; Yan, C.; Hu, A.; Yang, Y.T.; Chen, T.Y. Programming ORR Activity of Ni/NiO$_x$@Pd Electrocatalysts via Controlling Depth of Surface-Decorated Atomic Pt Clusters. *ACS Omega* **2018**, *3*, 8733–8744. [CrossRef]
34. Bhalothia, D.; Lin, C.-Y.; Yan, C.; Yang, Y.-T.; Chen, T.-Y. Effects of Pt metal loading on the atomic restructure and oxygen reduction reaction performance of Pt-cluster decorated Cu@Pd electrocatalysts. *Sustain. Energy Fuels* **2019**. [CrossRef]
35. Dai, S.; Chou, J.-P.; Wang, K.-W.; Hsu, Y.-Y.; Hu, A.; Pan, X.; Chen, T.-Y. Platinum-trimer decorated cobalt-palladium core-shell nanocatalyst with promising performance for oxygen-reduction reaction. *Nat. Commun.* **2019**, *10*, 440. [CrossRef] [PubMed]

© 2019 by the authors. Licensee MDPI, Basel, Switzerland. This article is an open access article distributed under the terms and conditions of the Creative Commons Attribution (CC BY) license (http://creativecommons.org/licenses/by/4.0/).

Article

Resistive Switching Characteristics of HfO$_2$ Thin Films on Mica Substrates Prepared by Sol-Gel Process

Chao-Feng Liu, Xin-Gui Tang *, Lun-Quan Wang, Hui Tang, Yan-Ping Jiang, Qiu-Xiang Liu, Wen-Hua Li and Zhen-Hua Tang

School of Physics & Optoelectric Engineering, Guangdong University of Technology, Guangzhou Higher Education Mega Center, Guangzhou 510006, China
* Correspondence: xgtang@gdut.edu.cn; Tel.: +86-20-3932-2265

Received: 15 July 2019; Accepted: 2 August 2019; Published: 4 August 2019

Abstract: The resistive switching (RS) characteristics of flexible films deposited on mica substrates have rarely been reported upon, especially flexible HfO$_2$ films. A novel flexible Au/HfO$_2$/Pt/mica resistive random access memory device was prepared by a sol-gel process, and a Au/HfO$_2$/Pt/Ti/SiO$_2$/Si (100) device was also prepared for comparison. The HfO$_2$ thin films were grown into the monoclinic phase by the proper annealing process at 700 °C, demonstrated by grazing-incidence X-ray diffraction patterns. The ratio of high/low resistance (off/on) reached 1000 and 50 for the two devices, respectively, being relatively stable for the former but not for the latter. The great difference in ratios for the two devices may have been caused by different concentrations of the oxygen defect obtained by the X-ray photoelectron spectroscopy spectra indicating composition and chemical state of the HfO$_2$ thin films. The conduction mechanism was dominated by Ohm's law in the low resistance state, while in high resistance state, Ohmic conduction, space charge limited conduction (SCLC), and trap-filled SCLC conducted together.

Keywords: resistance switching; high/low resistance; oxygen defect; conduction mechanism

1. Introduction

Resistive random access memory (RRAM) is a kind of memory in which, according to the different voltage applied to the metal oxide, the resistance of the material changes correspondingly between the high resistance state (HRS) and the low resistance state (LRS), so as to open or block the current flow channel and use this property to store various information [1]. RRAM can significantly increase durability and data transmission speed compared with flash memory devices. The main factor affecting the performance of RRAM is the RS layer, and the performance of different RS layers varies greatly. A variety of materials can be applied as the resistive switching layers of RRAM, such as HfO$_2$, SnWO$_4$, ZrO$_2$, and CuO [2–6], among which binary metal oxides like HfO$_2$ are widely regarded as the most promising resistive switching layer [1,7]. The conduction mechanisms of RRAM have been studied in depth, among which Ohmic conduction, Schottky emission, space-charge-limited conduction (SCLC), and trap-assisted tunneling are the most popular [1,8–11]. The conductive filament (CF) model has also been one of the most recognized models [8]. With the development of science and technology, flexible memory has also been extensively studied in the past decade [12,13]. Due to the advantages of their being inexpensive and lightweight, flexible memristors are more widely used than non-flexible devices such as disposable sensors [14] or indenofluorene-based monomers [15].

Although flexible electronic devices have promising applications in wearable devices, few papers have reported on the RS characteristics of flexible films deposited on mica substrates [16–18]. Mica substrates are cheap, easy to prepare, and satisfy the demands of industrial production, which makes them an excellent candidate for preparing flexible RRAM substrates. In this paper, HfO$_2$ thin films were grown on flexible mica substrates by the sol-gel method. For comparison of different substrates, HfO_2 films were also deposited on Pt/Ti/SiO$_2$/Si (100) substrates. As a kind of ordinary semiconductor compound, HfO_2 film has a high dielectric constant and desirable light transmittance with a simple preparation [19]. Due to its thermal stability and excellent retention performance [2,20–22], HfO_2 has been widely studied in the field of RRAM in recent years [23], and is one of the most promising candidates for the resistive switching layer. The results show that the ratio of HRS to LRS exceeded 100 in the HfO_2-based-non-flexible structure, with excellent stability. In contrast to non-flexible resistive switching, the HfO_2-based flexible structure demonstrated a pretty good resistive switching characteristic, but its endurance was inferior to non-flexible resistive switching. This HfO_2-based flexible device has a simple preparation method (sol-gel), inexpensive cost, and excellent flexibility not existing in an HfO_2-based-non-flexible structure, which conforms to the developing requirements of our time for flexible RRAM.

2. Materials and Methods

Using the sol-gel method for coating, a certain amount of hafnium acetone was weighed as the raw material, the magnetic stirrer was used to dissolve it in acetic acid until a colloid formed, the hafnium acetone colloid was spirally coated onto the different substrate by a rotary coating machine and then placed on a drying platform. The drying platform was heated from room temperature to 300 °C for 10 min, which decomposed hafnium acetone into HfO_2 at high temperature. In this paper, there were two samples of different substrates, HfO_2/Pt/Ti/SiO$_2$/Si and HfO_2/Pt/mica flexible structures. For further discussion, the structures of HfO_2/Pt/Ti/SiO$_2$/Si and HfO_2/Pt/mica are abbreviated as S1 and S2, respectively, as shown in Figure 1. Both S1 and S2 were annealed at 700 °C in air atmosphere for 30 min. After annealing, an Au point electrode with diameter of 0.5 mm was plated on the sample using a small high-vacuum coating machine and a mask template with diameter of 0.5 mm at room temperature for two min to form a top–bottom (TB) electrode structure.

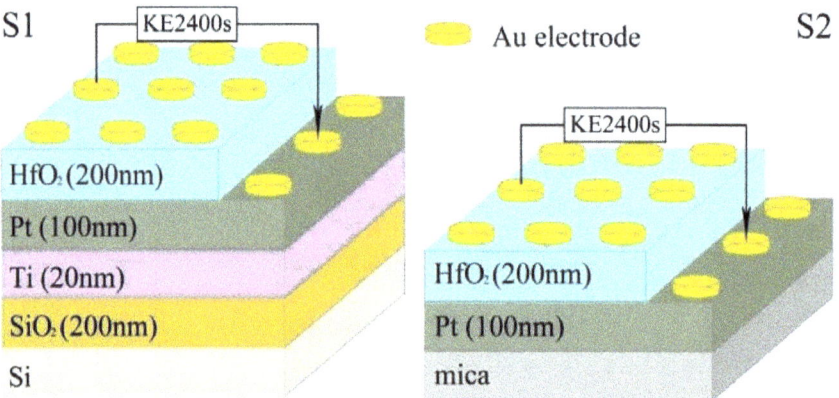

Figure 1. Schematic patterns of the HfO_2/Pt/Ti/SiO$_2$/Si (S1) and HfO_2/Pt/mica (S2) devices.

Current–voltage (I-V) and endurance characteristics were measured by the Keithley 2400 s instrument. Atomic force microscopy (AFM) showed the surface morphology of the film, and field emission scanning electron microscopy (FESEM) could clearly observe the thickness of the HfO_2 thin film and the layers between substrate and film. Additionally, the phase structures of HfO_2 films were analyzed by grazing-incidence X-ray diffraction (GIXRD) with an incident angle of 1°. Moreover, X-ray photoelectron spectroscopy (XPS) analyses of the HfO_2 thin films were carried out using an Escalab 250Xi X-ray photoelectron spectrometer.

3. Results and Discussion

It can be seen from Figure 2a,b that the grain size of the HfO$_2$ thin films after annealing was relatively small, which was due to the low annealing temperature and short annealing time. The SEM cross-sectional views of S1 and S2 show a dense layer of HfO$_2$ with a thickness of ~200 nm, and a dense Pt layer with a thickness of ~100 nm can be seen in all cases, as shown in Figure 2c,d. Additionally, the density and adhesion of HfO$_2$ on a typical Pt substrate were better than that on a flexible substrate. Figure 3 indicates the GIXRD patterns of the HfO$_2$ films grown on two different devices. As can be seen from Figure 3, the HfO$_2$ thin films had high crystallinity—a polycrystalline (100), (110), ($\bar{1}$11), (111), (200), and (220) oriented monoclinic phase structure [24,25]. Additionally, the PDF#78-0050 of the HfO$_2$ monoclinic phase is inserted in Figure 3 to better identify the XRD peak of the HfO$_2$ film. HfO$_2$ with a monoclinic phase structure can accumulate oxygen vacancies [26]. The relatively small GIXRD peak intensity shows the smaller grain size of the HfO$_2$ thin films, corresponding to the results of the SEM and AFM analyses. Additionally, a Pt (111) oriented peak existed in the S1 device.

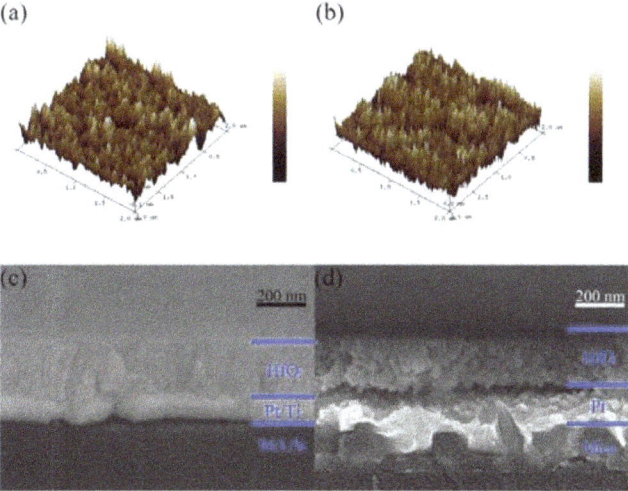

Figure 2. (**a**,**b**) Atomic force microscopy patterns of HfO$_2$ thin films in S1 and S2 devices, respectively; (**c**,**d**) typical cross-sectional scanning electron microscope images of S1 and S2, respectively.

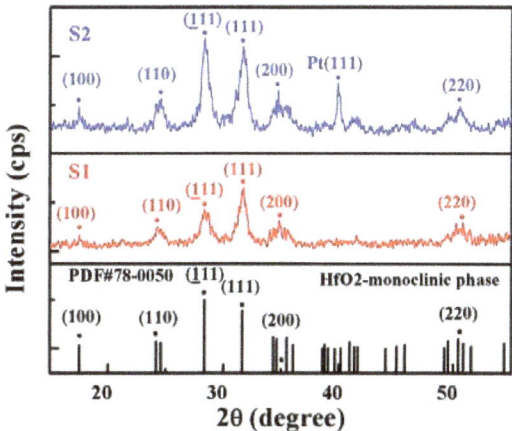

Figure 3. Grazing-incidence X-ray diffraction patterns of HfO$_2$ thin films in S1 and S2.

As shown in Figure 4a,c, Hf 4f core levels of HfO$_2$ thin films layers in all cases were deconvoluted into two Gaussian peaks (16.7 eV for Hf 4f5/2 and 18.3 eV for Hf 4f7/2, indicated by the red line and green line, respectively) [27–29]. Figure 4b,d shows XPS spectra of the O 1 s core levels of the HfO$_2$ thin films layers in all cases. Obviously, the Gaussian peak with a binding energy of 529.7eV was defined as lattice oxygen (O$_l$), corresponding to the oxygen in the HfO$_2$ matrix; the other, with a binding energy of 531.5eV, was defined as defect oxygen (O$_d$), caused by the defects of oxygen vacancies in the HfO$_2$ thin film layers. Previous research has indicated the higher the intensity of O$_d$, the higher the concentration of oxygen vacancy [5]. The ratio of Hf/O$_l$ in all devices was ~2, signifying the existence of HfO$_2$ [30,31]. Furthermore, the ratio of O$_l$/O$_d$ in S1 devices (0.32) was larger than that of S2 devices (0.25) and the ratio of O$_d$ in S1 devices to that in S2 devices was 0.82, resulting in the difference of HRS/LRS ratio between the two devices, which was consistent with I-V characteristics.

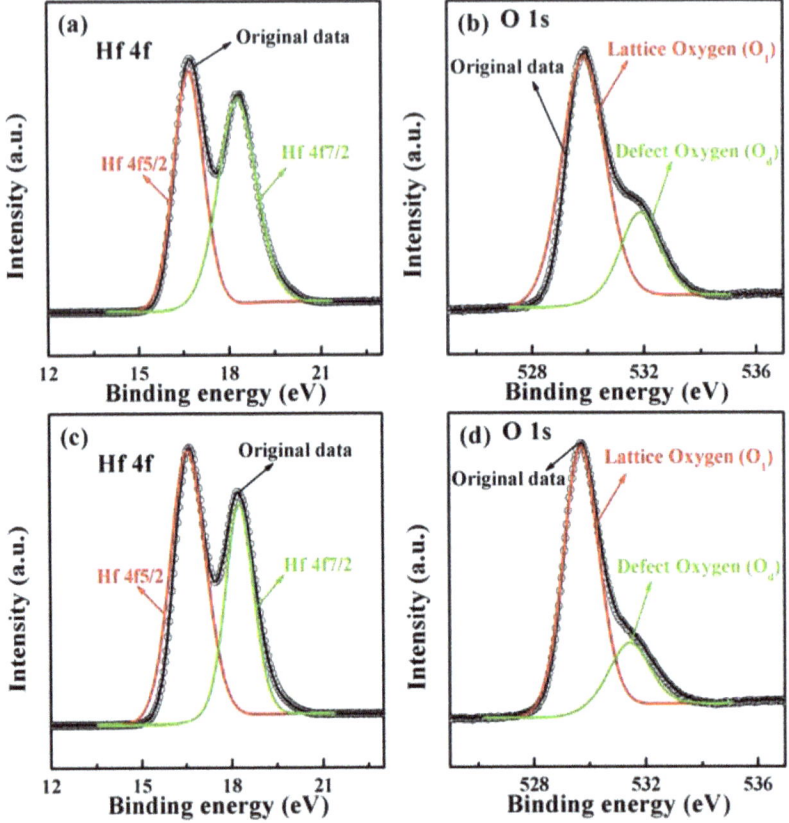

Figure 4. (a,b) The X-ray photoelectron spectroscopy (XPS) spectra of the S1 device; (c,d) the XPS spectra of the S2 device; (b,d) show the different oxygen intensities after fitting the peak.

Figure 5a,b shows the excellent resistance switching behaviors of the S1 and S2 structures. It is apparent that the V$_{set}$ and V$_{reset}$ of the S1 devices were 0.7 V and −0.5 V respectively, while the V$_{set}$ and V$_{reset}$ of the S2 devices were 0.7 V and −0.7 V respectively [22,32]. In addition, because the grain size of HfO$_2$ for S2 is larger than that for S1, based on the FESEM patterns (Figures S2 and S3), the switching currents of the S2 device were much larger than those of the S1 device. When the applied bias increased from 0 V to 0.7 V, both devices remain "off" (HRS). The device will be converted to LRS if the voltage reaches 0.7 V (V$_{set}$). Subsequently, with a voltage loop of 0.7 V to 1 V to −0.7 V for S2 (0.7 V to 1 V to

−0.5 V for S1), the device will always stay in "on" (LRS). When the voltage reaches −0.7 V (−0.5 V for S1) for the first time, the device will immediately be reset to "off" (HRS), and remain HRS all the way up to 0 V. The turn-on slope of S1 was calculated as 0.3 V/decade and was almost equal to that of S2, which depicted a switching speed in S2 consistent with S1; the ratio of HRS and LRS for the S1 device (~100) was greater than that of S2 device (~50), which also indicates that the S1 device had better switching characteristics than the S2 device. Additionally, resistive switching characteristics with 100 sweep cycles are depicted in Figure 2c,d. It can be seen clearly that the HRS/LRS ratio of S2 device gradually decreased from the 50th cycle; by contrast, the HRS/LRS ratio of the S1 device was almost stable when a forward bias was applied. From the results above, the device formed on the flexible substrate had the characteristics of typical RRAM. Figure 6 shows a stable resistance state (LRS/HRS) of the S1 device, with a reading voltage of 0.2 V for 100 sweep cycles at room temperature. The fitting linear curves in Figure 6a exhibit a stable off/on ratio for S1 RRAM devices, starting at 1000 times, slowly falling to 100 times, and then leveling off. However, as can be seen from Figure 6b, the S2 devices exhibited poor endurance characteristics, with rapid fatigue from 50 times to 10 times followed by leveling off. For the sake of illustrating the variation in HRS resistance and LRS resistance, Figure 6c,d compares the cumulative probability plots of HRS and LRS for the two devices at a reading voltage of 0.2 V. Compared to the S2 device, the S1 device exhibited a stable distribution of off/on resistance [33]. From the above analysis, the performance of S2 device was not as good as that of the S1device. In order to better illustrate the poor fatigue characteristics of S1 devices, repeatability tests are also conducted, as is shown in Figure S1. This demonstrates the shortcoming of mica-based devices that must be improved upon but cannot be at present.

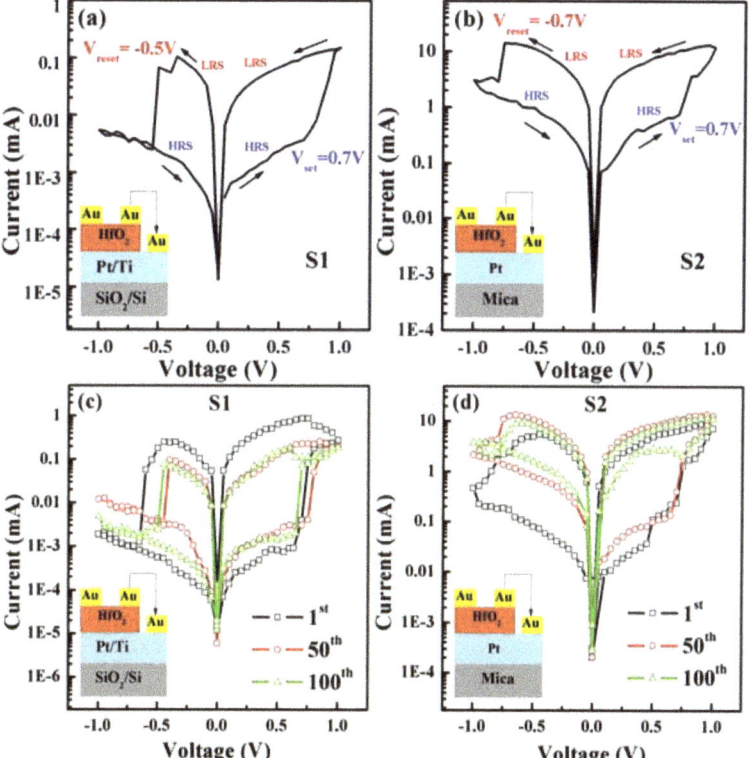

Figure 5. Resistive switching characteristics of (a) Au/HfO$_2$/Pt/Ti/SiO$_2$/Si, (b) Au/HfO$_2$/Pt/mica, (c) Au/HfO$_2$/Pt/Ti/SiO$_2$/Si with 100 sweep cycles, and (d) Au/HfO$_2$/Pt/mica with 100 sweep cycles.

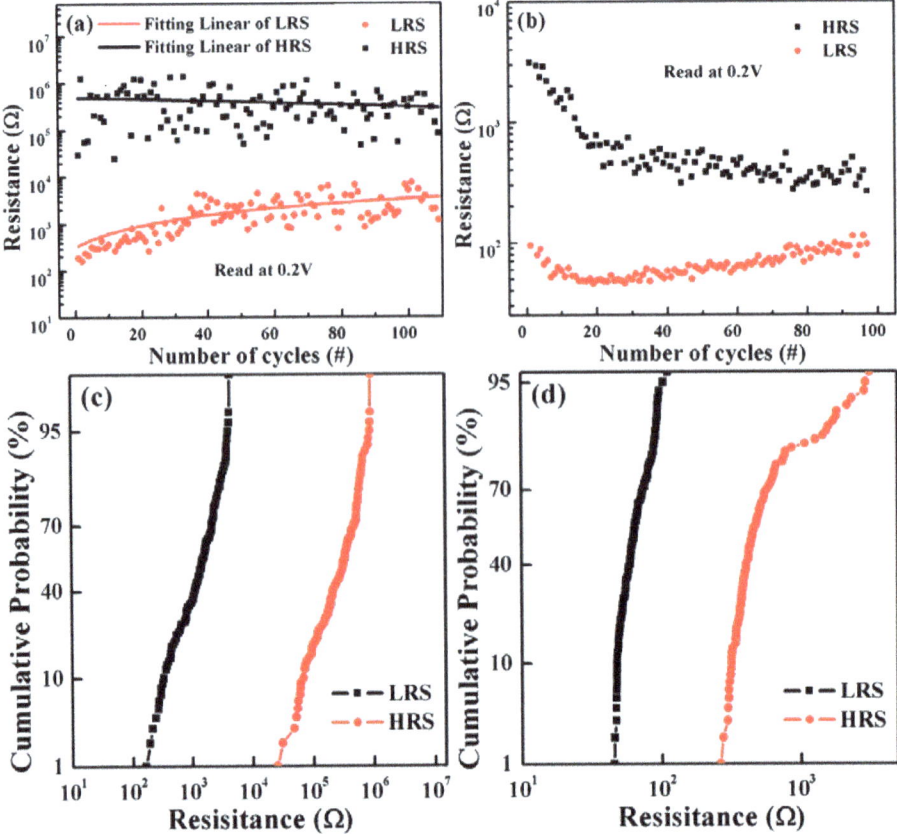

Figure 6. Endurance characteristics of (a) Au/HfO$_2$/Pt/Ti/SiO$_2$/Si and (b) Au/HfO$_2$/Pt/mica RRAM devices at room temperature; (c,d) the cumulative probability plots of high resistance state and low resistance state for the two devices, respectively, at a reading voltage of 0.2 V.

Figure 7 indicates that Ohmic conduction (I is proportional to V) and SCLC (I is proportional to V^2) were the main conduction mechanisms. The current density of SCLC can be depicted as following [1]:

$$J_{SCLC} = \frac{9}{8}\mu\varepsilon\frac{V^2}{d^3} \tag{1}$$

where ε is the permittivity of the film, μ is the electron mobility, V is the voltage, and d is the thickness of the film. Furthermore, it can be reasonably inferred that the conductive mechanism is dominated by trap-filled SCLC (indicated by the green line) when the forward bias is more than 0.7 V. The current density of trap-filled SCLC can be depicted as following [1]:

$$J_{TFSCLC} = q^{1-l}\mu N\left(\frac{2l+1}{l+1}\right)^{l+1}\left(\frac{l}{l+1}\frac{\varepsilon_r\varepsilon_0}{N_t}\right)^l\frac{V^{l+1}}{d^{2l+1}} \tag{2}$$

where q, l, μ, ε_r, ε_0, N_t, N, V, and d are the elemental charge, the ratio of the characteristic temperature of the trap distribution to the operating temperature, the carrier mobility, the permittivity of the film, the permittivity of free space, the trap density, the density of state in the conduction band or valence band, the applied voltage, and the film thickness, respectively.

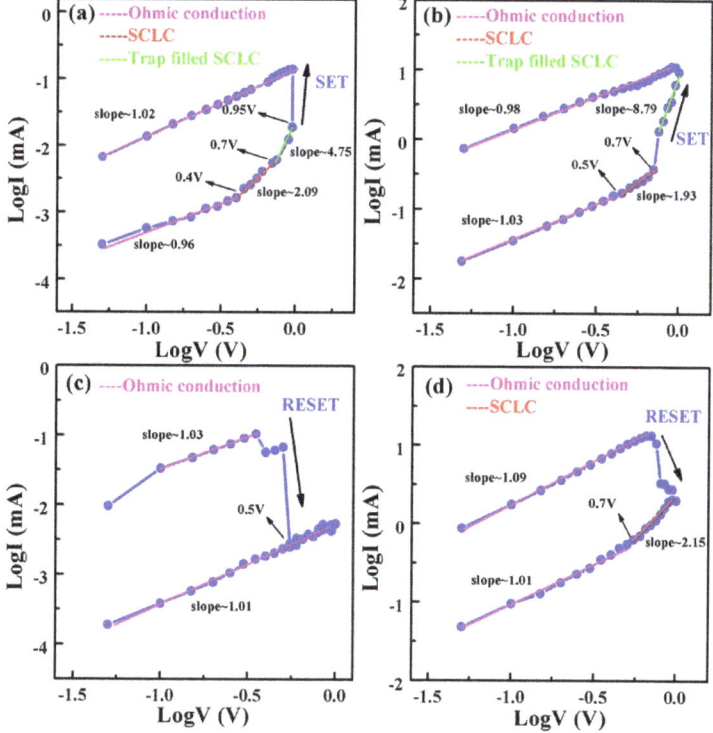

Figure 7. log$I$–log$V$ plots in (**a**) Au/HfO$_2$/Pt/Ti/SiO$_2$/Si and (**b**) Au/HfO$_2$/Pt/mica RRAM devices under positive voltage; (**c**) Au/HfO$_2$/Pt/Ti/SiO$_2$/Si and (**d**) Au/HfO$_2$/Pt/mica RRAM devices under negative voltage.

The log$I$ versus log$V$ plots have been fitted linearly to analyze the conduction mechanisms of S1 and S2 devices comprehensively. Figure 7a,b exhibits four different slope regions for S1 and S2 devices in positive sweeps, which represent three different conduction mechanisms: Ohmic conduction (slope = 1), SCLC (slope = 2), and trap-filled SCLC (slope > 2). The conduction mechanism of the S1 device was consistent with S2 device, which transferred from Ohmic conduction to SCLC at 0.4 V for the S1 device and 0.5 V for the S2 device, and then to trap-filled SCLC at 0.7 V for all cases. According to the SCLC mechanism, the electron trap is conceived as an oxygen vacancy, and the resistance slowly decreases as the oxygen vacancy filled with electrons, according to Child's law. However, when the oxygen vacancy is brimming with electrons, the latter will flow past the conduction band, so that the devices will be switched from HRS to LRS [34]. Note that the slope of LRS was almost equal to 1 for all devices, indicating the formation of CF. For the S1 devices in negative sweeps, the Ohmic mechanism ran through the LRS and HRS, as is shown in Figure 7c, while for S2 devices in negative sweeps, it can be clearly observed that the slope was 2.15 for voltage ranges from −1 V to −0.7 V, demonstrating that the CF formed by oxygen vacancies was broken, resulting in reset of resistance state from LRS to HRS. At the same time, the electrons were quickly disengaged from the oxygen vacancy. In conclusion, the conduction mechanism was dominated by Ohmic conduction in LRS, while in HRS, the Ohmic conduction and SCLC conducted together.

According to the analysis of XPS spectra and conduction mechanism, the CF caused by oxygen vacancy dominated the resistance switching mechanism [5,35]. As shown in Figure 8, a typical CF model has been proposed to better illustrate the influence of O$_d$. A large number of defects caused by oxygen vacancies exist in HfO$_2$ thin film layers, distributing randomly in the thin film layer and the interface layer without biased voltage, corresponding to the HRS depicted in Figure 8a, which is consistent with the HRS at zero voltage shown in Figure 5a,b. When a forward bias (<0.4 V for S1 devices, <0.5 V for S2 devices) was applied to the device, the conduction mechanism obeyed Ohm's law. The trap was gradually filled by injected electrons as the applied voltage increased (0.4 V–0.7 V for S1 devices, 0.5–0.7 V for S2 devices), the CF formed, as shown in Figure 8b, and the conduction mechanism was dominated by Child's law (SCLC). At this time, it corresponded to the HRS of the positive bias voltage (0–0.7 V) in Figure 5a,b. Due to the action of the electric field force, the oxygen ions drifted upward and accumulated at one end of the top electrode, forming a conductive bridge via these oxygen vacancies, while the CF built by oxygen vacancies connected the top and bottom electrodes, resulting in the SET process, as shown in Figure 8c [32,36]. It can also be seen from Figure 5a,b that when the forward voltage was greater than 0.7 V for the first time, the CF was formed, and the RS converted from HRS to LRS. When the voltage loop dropped from 1 V to −0.5 V, the RS remained "on" (LRS), as shown in Figure 5a,b, which is consistent with Figure 8c. Meanwhile, the conduction mechanism was controlled by Ohmic conduction for the existence of CF. Figure 8d exhibits that as the reverse bias was applied to the device, the oxygen ions drifted downward and then combined with the oxygen vacancy, resulting in the rupture of the CF. Combined with the analysis in Figure 5a,b, when the reverse bias voltage reached a certain value (−0.5 V for S1, −0.7 V for S2), the CF completely ruptured, resulting in an instant reset from LRS to HRS. Subsequently, the RS was always off (HRS) while the voltage loop went from −0.7 V to −1 V to 0 V for S2 or from −0.5 V to −1 V to 0 V for S1. The formation and rupture of the CF perfectly explains the principle of resistance switching, which is consistent with the conductive mechanism and I-V characteristics.

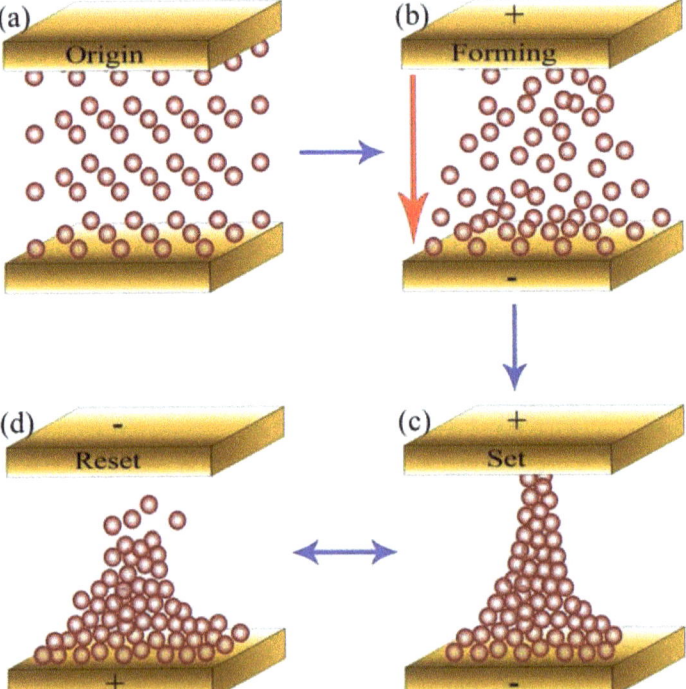

Figure 8. Schematic diagram explaining the conduction mechanism: (**a**) The RS is very high because the device does not form CF; (**b**) When a positive bias is applied, the oxygen vacancies move towards the negative electrode and a CF is formed; (**c**) The device is in the SET state because the oxygen vacancies has formed CF; (**d**) When the voltage is reversed, the CF immediately rupture.

4. Conclusions

In summary, an Au/HfO$_2$/Pt/Ti/SiO$_2$/Si device and an Au/HfO$_2$/Pt/mica device were fabricated by the sol-gel method. As a popular research material, the S1 device structure has been thoroughly studied. At present, the breakthrough point was whether the HfO$_2$ with a flexible structure would have the same performance as the typical device. Herein, quite a few advantages and disadvantages of flexible HfO$_2$ devices have been identified by analyzing the differences between the S1 and S2 devices. The O$_d$ intensity of XPS spectra for the S2 device was lower than for the S1 device, which indirectly illustrates that the HRS/LRS ratio of the S2 device was lower. Meanwhile, the *I-V* characteristic also demonstrated the difference in off/on ratio. Nevertheless, HRS/LRS ratio of the S2 device also reached 50, which is enough to illustrate the potential application of flexible HfO$_2$ device and that they are worth further study. For the Au/HfO$_2$/Pt/mica device, the conduction mechanism was dominated by Ohmic conduction in LRS, and Ohmic conduction and SCLC conduction together in HRS. There is no doubt that the CF model can perfectly illustrate this conduction mechanism. The potential problem is the poor fatigue characteristics of the HfO$_2$-mica-based RRAM, which cannot be solved at present, but we hope to solve effectively in the future.

Supplementary Materials: The following are available online at http://www.mdpi.com/2079-4991/9/8/1124/s1, Figure S1: Current–Voltage plots of repeated samples, Figure S2: The FESEM pattern of the surface for S1 device, Figure S3: The FESEM pattern of the surface for S2 device.

Author Contributions: Conceptualization, C.-F.L. and X.-G.T.; Data curation, C.-F.L.; Formal analysis, C.-F.L., L.-Q.W. and H.T.; Funding acquisition, X.-G.T.; Investigation, C.-F.L.; Methodology, X.-G.T.; Project administration, X.-G.T.; Resources, Y.-P.J., Q.-X.L., W.-H.L. and Z.-H.T.; Writing—original draft, C.-F.L.; Writing—review & editing, C.-F.L., X.-G.T., L.-Q.W. and H.T.

Funding: This research was funded by "the National Natural Science Foundation of China (Grant Nos. 11574057, 51604087and51702055)", "the Guangdong Provincial Natural Science Foundation of China (Grant No. 2016A030313718)", and "the Science and Technology Program of Guangdong Province of China (Grant Nos. 2016A010104018 and 2017A010104022)".

Conflicts of Interest: The authors declare no conflict of interest.

References

1. Lim, E.W.; Ismail, R. Conduction Mechanism of Valence Change Resistive Switching Memory A Survey. *Electronics* **2015**, *4*, 586–613. [CrossRef]
2. Lin, C.A.; Huang, C.J.; Tseng, T.Y. Impact of barrier layer on HfO$_2$-based conductive bridge random access memory. *Appl. Phys. Lett.* **2019**, *114*, 093105. [CrossRef]
3. Han, P.D.; Sun, B.; Cheng, S.; Yu, F.Y. An optoelectronic resistive switching memory behavior of Ag/α-SnWO$_4$/FTO device. *J. Alloy Compd.* **2016**, *681*, 516–521. [CrossRef]
4. Wei, X.D.; Huang, H.; Ye, C.; Wei, W.; Zhou, H.; Chen, Y.; Zhang, R.L.; Zhang, L.; Xia, Q. Exploring the role of nitrogen incorporation in ZrO$_2$ resistive switching film for enhancing the device performance. *J. Alloy. Compd.* **2019**, *775*, 1301–1306. [CrossRef]
5. Hsu, C.C.; Wang, T.C.; Tsao, C.C. Forming-free sol-gel ZrO$_x$ resistive switching memory. *J. Alloy. Compd.* **2018**, *769*, 65–70. [CrossRef]
6. Cheng, T.D.; Zhang, H.; Liu, N.; Yu, P.F.; Wu, C.T.; Tang, X.G. Improvement of memristive properties in CuO films with a seed Cu layer. *Appl. Phys. Lett.* **2019**, *114*, 061602.
7. Dai, Y.H.; Zhao, Y.Y.; Wang, J.Y.; Xu, J.B.; Yang, F. First principle simulations on the effects of oxygen vacancy in HfO$_2$-based RRAM. *AIP Adv.* **2015**, *5*, 017133. [CrossRef]
8. Xue, K.H.; Miao, X.S. Oxygen vacancy chain and conductive filament formation in hafnia. *J. Appl. Phys.* **2018**, *123*, 16150. [CrossRef]
9. Chen, C.; Pan, F.; Wang, Z.S.; Yang, J.; Zeng, F. Bipolar resistive switching with self-rectifying effects in Al/ZnO/Si structure. *Appl. Phys. Lett.* **2012**, *111*, 013702. [CrossRef]
10. Zhang, M.Y.; Long, S.B.; Li, Y.; Liu, Q.; Lv, H.B.; Miranda, E.A.; Sune, J.D.; Liu, M. Analysis on the Filament Structure Evolution in Reset Transition of Cu/HfO$_2$/Pt RRAM Device. *Nanoscale. Res. Lett.* **2016**, *11*, 269. [CrossRef]

11. Chand, U.; Huang, K.C.; Huang, C.Y.; Tseng, T.Y. Mechanism of Nonlinear Switching in HfO$_2$-Based Crossbar RRAM With Inserting Large Bandgap Tunneling Barrier Layer. *IEEE Trans. Electron Dev.* **2015**, *62*, 3665–3670. [CrossRef]
12. Gergel-Hackett, N.; Hamadani, B.; Dunlap, B.; Suehle, J.; Richter, C.; Hacker, C.; Gundlach, D. A Flexible Solution-Processed Memristor. *IEEE Electron Dev. Lett.* **2009**, *30*, 706–708. [CrossRef]
13. Zhou, Z.; Mao, H.; Wang, X.; Sun, T.; Chang, Q.; Chen, Y.; Xiu, F.; Liu, Z.; Liu, J.; Huang, W. Transient and flexible polymer memristors utilizing full-solution processed polymer nanocomposites. *Nanoscale* **2018**, *10*, 14824–14829. [CrossRef] [PubMed]
14. Moller, S.; Perlov, C.; Jackson, W.; Taussig, C.; Forrest, S.R. A polymer/semiconductor write-once read-many-times memory. *Lett. Nat.* **2003**, *426*, 166–169. [CrossRef] [PubMed]
15. Figà, V.; Ustab, H.; Macaluso, R.; Salznerd, U.; Ozdemirb, M.; Kulyke, B.; Krupkaf, O.; Bruno, M. Electrochemical polymerization of ambipolar carbonyl-functionalized indenofluorene with memristive properties. *Opt. Mater.* **2019**, *94*, 187–195. [CrossRef]
16. Gutowski, M.; Jaffe, J.E.; Liu, C.L.; Stoker, M.; Hegde, R.I.; Rai, R.S.; Tobin, P.J. Thermodynamic stability of high-K dielectric metal oxides ZrO$_2$ and HfO$_2$ in contact with Si and SiO$_2$. *Appl. Phys. Lett.* **2002**, *80*, 1897. [CrossRef]
17. Ku, B.; Abbas, Y.; Sokolov, A.S.; Choi, C. Interface engineering of ALD HfO$_2$-based RRAM with Ar plasma treatment for reliable and uniform switching behaviors. *J. Alloy Compd.* **2017**, *735*, 1181–1188. [CrossRef]
18. Chen, Y.Y.; Goux, L.; Clima, S.; Govoreanu, B.; Degraeve, R.; Kar, G.S.; Fantini, A.; Groeseneken, G.; Wouters, D.J.; Jurczak, M. Endurance/retention trade-off on HfO$_2$/metal cap 1T1R bipolar RRAM. *IEEE Trans. Electron Dev.* **2013**, *60*, 1114–1121. [CrossRef]
19. Song, B.; Cao, R.; Xu, H.; Liu, S.; Liu, H.J.; Li, Q.J. A HfO$_2$/SiTe Based Dual-Layer Selector Device with Minor Threshold Voltage Variation. *Nanomaterials* **2019**, *9*, 408. [CrossRef]
20. Nguyen, T.H.; Barua, A.; Bailey, T.; Rush, A.; Kosel, P.; Leedy, K.; Jha, R. Reflection coefficient of HfO$_2$-based RRAM in different resistance states. *Appl. Phys. Lett.* **2018**, *113*, 192101. [CrossRef]
21. Yang, Y.X.; Yuan, G.L.; Yan, Z.B.; Wang, Y.J.; Lu, X.B.; Liu, J.M. Flexible Semitransparent, and Inorganic Resistive Memory based on BaTi$_{0.95}$Co$_{0.05}$O$_3$ Film. *Adv. Mater.* **2017**, *29*, 1700425. [CrossRef]
22. Xiao, Z.A.; Zhao, J.H.; Lu, C.; Zhou, Z.Y.; Wang, H.; Zhang, L.; Wang, J.J.; Li, X.Y.; Wang, K.Y.; Zhao, Q.L.; et al. Characteristic investigation of a flexible resistive memory based on a tunneling junction of Pd/BTO/LSMO on mica substrate. *Appl. Phys. Lett.* **2018**, *113*, 223501. [CrossRef]
23. Yang, C.H.; Han, Y.J.; Qian, J.; Lv, P.P.; Lin, X.J.; Huang, S.F.; Cheng, Z.X. Flexible Temperature-Resistant, and Fatigue-Free Ferroelectric Memory Based on Bi(Fe$_{0.93}$Mn$_{0.05}$Ti$_{0.02}$)O$_3$ Thin Film. *ACS Appl. Mater. Interfaces* **2019**, *11*, 12647–12655. [CrossRef]
24. He, G.; Liu, M.; Zhu, L.Q.; Chang, M.; Fang, Q.; Zhang, L.D. Effect of post deposition annealing on the thermal stability and structural characteristics of sputtered HfO$_2$ films on Si (100). *Surf. Sci.* **2014**, *576*, 67–75. [CrossRef]
25. Kumar, S.; Rai, S.B.; Rath, C. Latent Fingerprint Imaging Using Dy and Sm Codoped HfO$_2$ Nanophosphors: Structure and Luminescence Properties. *Part. Part. Syst. Charact.* **2019**, *36*, 1900048. [CrossRef]
26. Bradley, S.R.; Bersuker, G.; Shluger, A.L. Modelling of oxygen vacancy aggregates in monoclinic HfO$_2$: Can they contribute to conductive filament formation? *J. Phys. Condens. Matter.* **2015**, *27*, 415401. [CrossRef]
27. Crist, B.V. Handbook of Monochromatic XPS Spectra, Semiconductors. *IEEE Electr. Insul. M.* **2003**, *19*, 47.
28. Yoon, J.H.; Song, S.J.; Yoo, I.; Seok, J.Y.; Yoon, K.J.; Kwon, D.E.; Park, T.H.; Hwang, C.S. Highly Uniform, Electroforming-Free, and Self-Rectifying Resistive Memory in the Pt/Ta$_2$O$_5$/HfO$_{2-x}$/TiN Structure. *Adv. Funct. Mater.* **2014**, *24*, 5086–5095. [CrossRef]
29. Ma, H.L.; Zhang, X.M.; Wu, F.C.; Luo, Q.; Gong, T.C.; Yuan, P.; Xu, X.X.; Liu, Y.; Zhao, S.J.; Zhang, K.P.; et al. Self-Rectifying Resistive Switching Device Based on HfO$_2$/TaO$_x$ Bilayer Structure. *IEEE Trans. Electron Dev.* **2019**, *66*, 924–928. [CrossRef]
30. Januar, M.; Prakoso, S.P.; Lan, S.Y.; Mahanty, R.K.; Kuo, S.Y.; Liu, K.C. The role of oxygen plasma in the formation of oxygen defects in HfO$_x$ films deposited at room temperature. *J. Mater. Chem. C* **2015**, *3*, 4104–4114. [CrossRef]
31. Sokolov, A.S.; Jeon, Y.R.; Kim, S.; Ku, B.; Lim, D.; Han, H.; Chae, M.G.; Lee, J.; Ha, B.G.; Choi, C. Influence of oxygen vacancies in ALD HfO$_{2-x}$ thin films on non-volatile resistive switching phenomena with a Ti/HfO$_{2-x}$/Pt structure. *Appl. Surf. Sci.* **2018**, *434*, 822–830. [CrossRef]

32. Traore, B.; Blaise, P.; Vianello, E.; Grampeix, H.; Jeannot, S.; Perniola, L.; De Salvo, B.; Nishi, Y. On the Origin of Low-Resistance State Retention Failure in HfO$_2$-Based RRAM and Impact of Doping/Alloying. *IEEE Trans. Electron Dev.* **2015**, *62*, 4029–4036. [CrossRef]
33. Januar, M.; Prakoso, S.P.; Lan, S.Y.; Mahanty, R.K.; Kuo, S.Y.; Liu, K.C. Metal oxide resistive memory switching mechanism based on conductive filament properties. *J. Appl. Phys.* **2011**, *110*, 124518.
34. Padovani, A.; Larcher, L.; Pirrotta, O.; Vandelli, L.; Bersuker, G. Microscopic Modeling of HfO$_x$ RRAM Operations: From Forming to Switching. *IEEE Trans. Electron Dev.* **2015**, *62*, 1998–2006. [CrossRef]
35. Chen, P.H.; Su, Y.T.; Chang, F.C. Stabilizing Resistive Switching Characteristics by Inserting Indium-Tin-Oxide Layer as Oxygen Ion Reservoir in HfO$_2$-Based Resistive Random Access Memory. *IEEE Trans. Electron Dev.* **2019**, *66*, 1276–1280. [CrossRef]
36. Qi, M.; Tao, Y.; Wang, Z.Q.; Xu, H.Y.; Zhao, X.N.; Liu, W.Z.; Ma, J.G.; Liu, Y.C. Highly uniform switching of HfO$_{2-x}$ based RRAM achieved through Ar plasma treatment for low power and multilevel storage. *Appl. Surf. Sci.* **2018**, *458*, 216–221. [CrossRef]

© 2019 by the authors. Licensee MDPI, Basel, Switzerland. This article is an open access article distributed under the terms and conditions of the Creative Commons Attribution (CC BY) license (http://creativecommons.org/licenses/by/4.0/).

Article

Tailoring IGZO Composition for Enhanced Fully Solution-Based Thin Film Transistors

Marco Moreira, Emanuel Carlos, Carlos Dias, Jonas Deuermeier, Maria Pereira, Pedro Barquinha *, Rita Branquinho *, Rodrigo Martins and Elvira Fortunato

i3N/CENIMAT, Department of Materials Science, Faculty of Science and Technology, Universidade NOVA de Lisboa and CEMOP/UNINOVA, Campus de Caparica, 2829-516 Caparica, Portugal
* Correspondence: pmcb@fct.unl.pt (P.B.); ritasba@fct.unl.pt (R.B.)

Received: 4 August 2019; Accepted: 3 September 2019; Published: 6 September 2019

Abstract: Solution-processed metal oxides have been investigated as an alternative to vacuum-based oxides to implement low-cost, high-performance electronic devices on flexible transparent substrates. However, their electrical properties need to be enhanced to apply at industrial scale. Amorphous indium-gallium-zinc oxide (a-IGZO) is the most-used transparent semiconductor metal oxide as an active channel layer in thin-film transistors (TFTs), due to its superior electrical properties. The present work evaluates the influence of composition, thickness and ageing on the electrical properties of solution a-IGZO TFTs, using solution combustion synthesis method, with urea as fuel. After optimizing the semiconductor properties, low-voltage TFTs were obtained by implementing a back-surface passivated 3-layer In:Ga:Zn 3:1:1 with a solution-processed high-κ dielectric; AlO$_x$. The devices show saturation mobility of 3.2 cm^2 V^{-1} s^{-1}, I_{On}/I_{Off} of 10^6, SS of 73 mV dec^{-1} and V_{On} of 0.18 V, thus demonstrating promising features for low-cost circuit applications.

Keywords: IGZO composition; solution combustion synthesis; transparent amorphous semiconductor oxides; low voltage operation

1. Introduction

In recent years, the emergence of flexible electronics has increased scientific interest in transparent amorphous metal oxide thin-film transistors (TFTs) deposited at low temperatures on flexible substrates. These devices are expected to meet technological demands for a wide range of flexible electronic concepts, such as foldable displays or signal readout/processing circuitry integrated in smart surfaces [1,2]. The advent of printed electronics research has led to the development of solution processed oxides deposited by techniques such as spin coating [3] and ink-jet printing [4], as an economic alternative to vacuum-based techniques [2]. In this regard, solution-processed amorphous oxide semiconductors TFTs offer low-cost, high-throughput and large-area scalability [5,6]. However, with sol-gel methods, it is difficult to modulate oxygen conditions that are crucial to form oxygen vacancies, which are a source for free carriers; therefore, electrical properties are controlled by post-annealing or composition of the metal oxide [7]. Up until now, a variety of solution-produced transparent oxides such as zinc oxide (ZnO) [8], zinc-tin oxide (ZTO) [9] and indium-zinc oxide (IZO) [10] have been a matter of study. Nevertheless, indium-gallium-zinc oxide (IGZO) remains the most used oxide semiconductor and Table 1 summarizes the reported properties of solution based IGZO TFTs produced with different processing conditions.

Table 1. Solution-based indium gallium zinc oxide (IGZO) thin film transistors produced at T ≥ 300 °C by spin-coating deposition of 2-methoxyethanol (2-ME) based precursors.

Year	Fuel	T_{max} (°C)	W/L	Dielectric (Technique)	In:Ga:Zn Ratio	μ_{SAT} (cm^2/Vs)	SS (V/dec)	I_{On}/I_{Off}	V_{on} (V)	V_{GS} (V)
2008 [11]		450	1000/150	SiN$_x$ (PECVD)	1:1:2	0.96	1.39	10^6	−5	−15 to 55
2009 [12]		400	1000/150	SiN$_x$ (PVD)	1:1:2	0.56 (μ_{ef})	2.81	4.6×10^6	5	−30 to 30
					3:1:2	0.90 (μ_{ef})	1.16	3.8×10^6	−0	
					5:1:2	1.25 (μ_{ef})	1.05	4.1×10^6	−10	
2009 [13]		400	100/50	SiO$_2$	2:1:2	2 (μ_{ef})	-	10^5	-	−40 to 40
2010 [14]	No	400	1000/90	ATO (ALD)	3:1:1	5.8 (μ_{lin})	0.28	6×10^7	−0	−10 to 30
2010 [15]		500	200/20	SiO$_2$ (Thermal oxidation)	4:1:5	1.13	2.5	-	-	−30 to 40
2010 [7]		450	1000/150	SiN$_x$	3:1:2	0.86 (μ_{lin})	0.63	10^6	−0	−30 to 30
2010 [16]		300	1000/100	SiO$_2$	63:10:27	0.2	-	10^5	~−15	−40 to 40
2011 [17]		300	500/100	SiO$_2$	5:1:2	0.003	2.39	4.5×10^4	-	−20 to 30
2013 [18]		300	1000/100	SiO$_2$ (Thermal oxidation)	62:5:23	1.73 (μ_{ef})	0.32	10^7	11	−10 to 40
2013 [19]	Yes (acac)	300	5000/100	SiO$_2$ (Thermal Oxidation)	80:10:10	5.43	-	10^8	-	0 to 100
2019 [20]		300	1000/100	SiO$_2$	10:1:3	1.62	0.03	10^6	−0	−40 to 80
2019 [3]	No	350	n.d./100	SiO$_2$	68:10:22	0.72	0.68	10^6	−0	−30 to 30
This work	Yes (urea)	300	160/20	AlO$_x$	3:1:1	3.2	0.073	10^6	0.18	−1 to 2

W/L: Width/Length; PECVD: plasma enhanced vapor deposition PVD: physical vapor deposition; ATO: aluminum-doped tin oxide; ALD: atomic layer deposition; acac: acetylacetone; n.d.: not defined.

The first reported IGZO solution TFTs were fabricated using high annealing temperatures (>400 °C) [11–13,15] in order to remove organic ligands groups from sol-gels, i.e., to convert completely metal-hydroxide (M–OH) species into metal-oxygen-metal (M–O–M). Nonetheless, high temperature annealing restricts the application of the films on most flexible polymeric substrates [21,22]. Some reports tested ~300 °C [17,18]; however, at lower temperatures, incomplete decomposition of the organic precursors might occur, and most of the M–OH species are not fully converted into M–O–M, severely affecting the semiconductor's electrical performance [18,23]. In 2011, Marks et al. [24] reported, for the first time, a novel method to produce thin films at lower temperatures: the combustion synthesis method. By introducing an oxidizing agent (metal nitrates) and a fuel as reducing agent into a precursor solution, the potential of the oxide precursor is enhanced; when the film is annealed at 200–300 °C, a local highly exothermic chemical reaction initiates within the film, forming M–O–M lattice where the applied temperature acts only as reaction initiator [24,25]. Acetylacetone [19,26] and urea [27,28] are the most commonly used fuels in this method for different solution-based semiconductors. IGZO is applied mainly as semiconducting n-channel layer in TFTs, due to its high field-effect mobility, small subthreshold slope (SS), stability and good uniformity [29–31]. In^{3+} cations are the main element of conduction band and due to the overlap of their 5s orbitals, IGZO exhibits high mobility, even in its amorphous form; Zn^{2+} contributes to stabilization and enhancement of electrical properties; and Ga^{3+} forms strong bonds with oxygen, controlling the carrier concentration so the material might act as a semiconductor, although this reduces the electron mobility compared to IZO [1,2,16]. Although there are a few reports regarding the effect of sol-gel IGZO composition on TFTs performance [14,15], the study on how electrical properties of solution combustion synthesis IGZO depend on material composition is still lacking. In this work, we discuss the influence of In:Ga:Zn cations ratios of combustion solution-processed IGZO, as well as the number of implemented layers on TFTs performance. Urea was chosen as fuel to use throughout this work, since it is more environment-friendly and less-expensive when compared to acetylacetone. Solution-based aluminum oxide (AlO_x) dielectric was implemented in IGZO TFTs, as superior device performance can be achieved by combining a high-κ oxide dielectric with a semiconductor material, namely, increased mobility and lower operation voltage compared to conventional SiO_2 dielectric [32].

2. Experimental Section

2.1. Precursor Solution Development and Characterization

The metallic oxide precursor solutions were prepared by individually dissolving indium (III) nitrate hydrate ($In(NO_3)_3 \cdot xH_2O$, Sigma, 99.9%, Darmstadt, Germany), gallium (III) nitrate hydrate ($Ga(NO_3)_3 \cdot xH_2O$, Sigma, 99.9%, Darmstadt, Germany), zinc nitrate hexahydrate ($Zn(NO_3)_2 \cdot 6H_2O$, ACROS Organics, 98%, Geel, Belgium) and aluminum nitrate non-hydrate ($Al(NO_3)_3 \cdot 9H_2O$, Carl Roth, ≥98%, Darmstadt, Germany) in 2-Methoxyethanol (2-ME) ($C_3H_8O_2$, ACROS Organics, >99.5%, Geel, Belgium), to yield solutions with a concentration of 0.2 M. Urea ($CO(NH_2)_2$, Sigma, 98%, Darmstadt, Germany) was added as fuel to each precursor solution for the combustion reaction, with molar ratios between urea and indium nitrate, gallium nitrate, zinc nitrate and aluminum nitrate of 2.5:1, 2.5:1, 1.67:1 and 2.5:1, respectively, to guarantee the redox stoichiometry of the reaction (see Tables S1–S5 in the Supplementary Materials). All precursor solutions were magnetically stirred at 430 rpm for 1 h at room temperature in air environment. IGZO precursor solutions were prepared by mixing indium nitrate, gallium nitrate and zinc nitrate precursor solutions to yield In:Ga:Zn molar ratios of 1:1:1, 2:1:1, 2:1:2 and 3:1:1, all with a 0.2 M concentration. IGZO and AlO_x solutions were magnetically stirred at 430 rpm for at least 24 h at room temperature in air environment. All solutions were filtrated through 0.2 μm hydrophilic filters. Precursor solutions viscosity measurements were performed in a BROOKFIELD Cap 2000+ (Brookfield Engineering Laboratories, Inc., Middleboro, MA, USA) using a Cap01 spindle at 30 °C with a 500 rpm speed.

Thermal and chemical characterization of precursor solutions were performed by differential scanning calorimetry (DSC) and thermogravimetry (TG) and Fourier transform-infrared spectroscopy (FTIR). DSC and TG analysis of dried precursor solutions were performed under air atmosphere up to 500 °C with a 10 °C min^{-1} heating rate in an aluminum crucible with a punctured lid using a simultaneous thermal analyzer, Netzsch (TG-DSC—STA 449 F3 Jupiter, Selb, Germany). FTIR spectroscopy characterization of IGZO solutions was performed using a Thermo Nicolet 6700 Spectrometer (Waltham, MA, USA) equipped with a single bounce diamond crystal Attenuated Total Reflectance (ATR) sampling accessory (Smart iTR). The spectra were acquired with a 4 cm^{-1} resolution in the range of 4000–525 cm^{-1} with a 45° incident angle.

2.2. IGZO Film Deposition and Material Characterization

Prior to deposition all substrates (p$^+$Si with a 100 nm thermally grown SiO$_2$ layer, Si wafer and soda-lime glass, 2.5 × 2.5 cm^2) were cleaned in an ultrasonic bath at 60 °C in acetone for 15 min, then in isopropyl alcohol (IPA) for 15 min. Subsequently, the substrates were cleaned with deionized water (DIW) and dried under N$_2$, followed by a 15 min ultraviolet (UV)/ozone surface activation step using a PSD-UV Novascan system (Ames, IA, USA). IGZO thin films were deposited onto SiO$_2$ substrates by sequentially spin coating one to three layers of IGZO precursor solution for 35 s at 3000 rpm (Laurell Technologies, North Wales, PA, USA), followed by an immediate hot plate annealing at 300 °C for 30 min in air after each layer to ensure the exothermic reaction. The AlO$_x$ dielectric precursor solution was spin coated at 2000 rpm for 35 s onto Si substrates and annealed at 300 °C for 30 min. FTIR spectroscopy characterization of thin films deposited on Si substrates was performed the same way as used for IGZO precursor solutions. The structure of the films was assessed by grazing angle X-Ray diffraction (GAXRD), using a X'Pert PRO PANalytical (Royston, UK) diffractometer with Cu Kα line radiation (λ = 1.540598 Å) and an incidence angle of the X-ray beam fixed at 0.75°, in the range of 20° to 50° (2θ). Surface morphology of the thin films was studied by scanning electron microscopy (SEM, Zeiss Auriga Crossbeam electron microscope) (Oberkochen, Germany) and atomic force microscopy (AFM, Asylum MFP3D, Asylum Research, Santa Barbara, CA, USA). Electron dispersive X-ray spectroscopy (EDS) was performed to study the chemical composition of the thin films. Optical characterization of the thin films was obtained with a Perkin Elmer lambda 950 UV/visible (Vis)/near infrared (NIR) (Llantrisant, UK) spectrophotometer, by measuring transmittance variation in a wavelength range from 200 to 2500 nm. Spectroscopic ellipsometry was used to measure thickness and band gap energy of thin films deposited on Si substrates, with an energy range from 1.5 to 5.5 eV and an incident angle of 45° using a Jobin Yvon Uvisel system (Chilly-Mazzarin, France). DELTAPSI software (v2.6.6.212, Horiba, Bensheim, Germany) was used to modulate the acquired data, and the fitting procedure was done pursuing the minimization of the error function (χ^2). X-ray photoelectron spectroscopy (XPS) of IGZO thin films was measured with a Kratos Axis Supra (Manchester, UK), using monochromated Al Kα irradiation (1486.6 eV). The detail spectra of the surfaces were acquired with an X-ray power of 225 W and a pass energy of 10 eV. Depth profiles were done using argon clusters of 500 atoms and 10 keV, scanned over an area of 1.5 × 1.5 mm^2 and a time per etch step of 100 s. The cluster mode was used in order to limit damage to the film introduced by the argon beam with respect to a conventional monoatomic mode. Here, the XPS acquisition parameters were 300 W and 40 eV pass energy, and an aperture was used to limit the measurement spot to 110 µm in diameter.

2.3. TFTs/Devices Fabrication and Characterisation

TFTs were produced in a staggered bottom-gate, top-contact structure by spin coating IGZO thin films onto 100-nm-thick thermal SiO$_2$ (C_i = 35 nF·cm^{-2}) or onto spin-coated 20-nm-thick AlO$_x$ (C_i = 306 ± 2 nF·cm^{-2}), both on Si wafers. Aluminum (Al) source and drain electrodes (80 nm thick) were deposited on IGZO films via a shadow mask by thermal evaporation, with channel width (W) and length (L) of 1400 µm and 100 µm, respectively (W/L = 14). A post-annealing step was performed on a hot plate for 1 h at 120 °C in air environment. Optimized IGZO/AlO$_x$ devices were patterned by standard photolithographic processes (W/L = 160/20) and passivated with 1 µm thick parylene. Electrical characterization was performed by measuring current-voltage characteristics of the devices using a semiconductor parameter analyzer (Agilent 4155C, Agilent Technologies, Santa Clara, CA, USA) attached to a microprobe station (Cascade M150, FormFactor, Livermore, CA, USA), inside a Faraday cage, in the dark and at room temperature.

Transfer curves were performed in double sweep mode and used to extract turn-on voltage (V_{On}), threshold voltage (V_T), hysteresis (V_{Hyst}), subthreshold slope (SS), mobility in saturation regime (μ_{Sat}) and on/off current ratio (I_{On}/I_{Off}). A gate-to-source voltage (V_{GS}) from −10 to 20 V and a drain-to-source voltage (V_{DS}) of 20 V were applied. SS was estimated by [33]:

$$SS = \left(\left.\frac{d\log(I_{DS})}{dV_{GS}}\right|_{max}\right)^{-1}$$

V_T was derived from a linear fitting $\sqrt{I_{DS}}$ vs. V_{GS} in the saturation region [33]:

$$I_{DS} = \frac{WC_i}{2L}\mu_{Sat}(V_{GS} - V_T)^2$$

where W and L are the channel width and length, and C_i is the gate dielectric capacitance per unit area. μ_{Sat} was obtained by [33]:

$$\mu_{Sat} = \left(\frac{\partial\sqrt{I_{DS}}}{\partial V_{GS}}\right)^2 \frac{2L}{WC_i}$$

Positive gate bias stress tests were performed on IGZO/AlO$_x$ TFTs using a semiconductor parameter analyzer (Keysight 4200SCS, Penang, Malaysia) and probe station (Janis ST-500, Woburn, MA, USA) under air environment by applying a constant gate voltage (0.5 MV·cm^{-1} electric field) for one hour.

3. Results and Discussion

3.1. Precursor Solutions and Thin Films Characterization

Solution combustion synthesis is an efficient method to obtain high-quality thin films at lower temperatures than sol-gel, by initiating an exothermic reaction between an oxidizer (usually nitrates) and an organic fuel acting as reducing agent. The generated localized energy efficiently converts the metal nitrates precursors into oxides [24,34].

The thermal characterization of the precursor solutions is relevant to evaluate the decomposition of metal oxides and the combustion reaction ignition temperature. Figure 1 illustrates DSC-TG data of IGZO 3:1:1 0.2 M solutions with and without urea.

The intense exothermic peaks, accompanied by a significant weight loss, correspond to the combustion reaction of residual fuel in the formation of IGZO thin films. For the precursor solution without urea, two exothermic peaks are observed at 110 °C and 340 °C, which can indicate the formation of two distinct materials due to a non-uniform distribution of the metal cations within the gel phase of the reaction. This has also been observed for the formation of other multicomponent oxides, such as ZTO, where the presence of more than one metal cation can lead to multistep synthesis and consequently ununiform material [35]. Thus, the complete conversion of the precursor without urea requires temperatures above 300 °C. When urea is used as fuel, only one exothermic peak is observed at 230 °C, which indicates that the IGZO formation occurs in one step, thus contributing to the uniform cation distribution. This conclusion is corroborated by comparison of In:Ga:Zn ratios by EDS and XPS analysis, as discussed further below (Figure 2b–d).

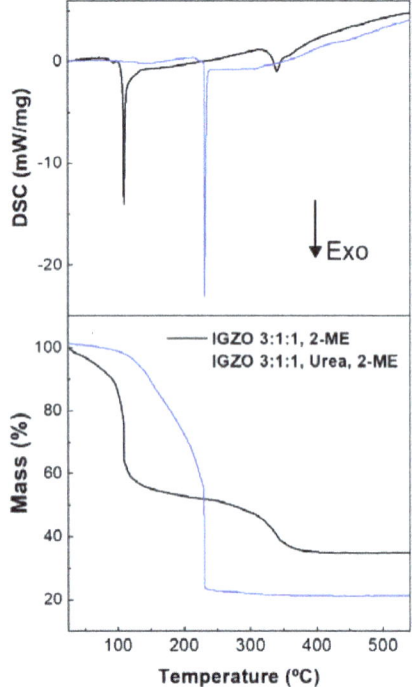

Figure 1. Differential scanning calorimetry (DSC)-thermogravimetry (TG) analysis of the IGZO (3:1:1) precursor solution with 2-ME as solvent and using with or without urea as fuel.

In the solution combustion precursor, no weight loss was observed above 230 °C, suggesting that this annealing temperature is enough to eliminate all organics in IGZO films. Nevertheless, for IGZO films deposition the annealing was perform at 300 °C in order to assure the complete formation of M–O–M. To study the influence of composition and thickness in produced IGZO films' properties, combustion precursor solutions were prepared with In:Ga:Zn ratio of 1:1:1; 2:1:1; 2:1:2 and 3:1:1 and 1-, 2- and 3-layered films IGZO thin films were deposited by spin-coating.

FTIR spectra of all IGZO films were performed after annealing at 300 °C and compared to spectra of precursor solutions which confirm the removal of organic compounds and the presence of M-O bonds in thin films after annealing (see Figure S1 and Table S6 in the Supplementary Materials).

IGZO films thickness was assessed by spectroscopic ellipsometry for all conditions. As expected, film thickness does not vary significantly for different composition as for all precursor solutions the viscosity is 2.30 ± 0.04 cP (Table S7), since concentration was maintained constant at 0.2 M. Additionally, film thickness increases almost linearly with the number of deposited layers with d ≈ 14 nm for 1-layer films; d ≈ 27 nm for 2-layer films and d ≈ 37 nm for 3-layer films (Table S8). Optical characterization shows typical average transmittance in the visible range of ~88% (Figure S2) as expected for IGZO thin films.

Optical bandgap energy (E_g) calculation by spectroscopic ellipsometry for combustion IGZO films with different In:Ga:Zn ratio (Table S9) reveals higher E_g for IGZO 1:1:1 (Ga-rich, E_g = 3.68 ± 0.03 eV), which is expected due to the higher E_g of GaO_x compared to InO_x and ZnO, whereas for the remaining compositions E_g = 3.45 ± 0.06 eV which is in agreement with reported values of IGZO films [29].

XRD, AFM and SEM analysis were performed to assess structural and morphological characteristics of the thin films. XRD analysis of solution processed IGZO was obtained by spin coating three layers on Si substrates. The absence of diffraction peaks indicates that no long-range order is present, as expected for multicomponent oxides, and amorphous films are obtained up to 300 °C (Figure 2a).

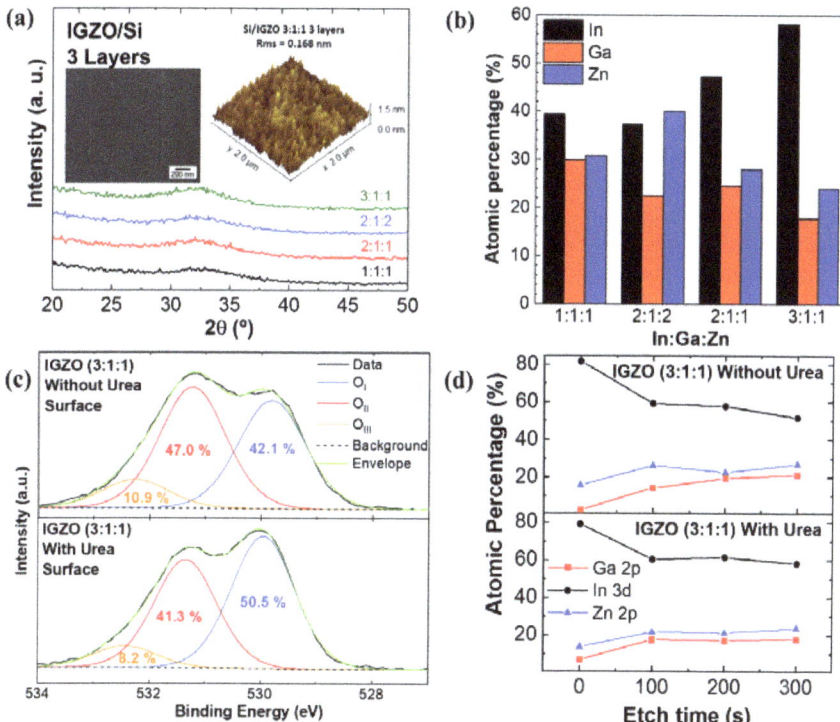

Figure 2. (a) X-Ray diffraction (XRD) of combustion IGZO films with different In:Ga:Zn ratio; inset shows the scanning electron microscopy (SEM) surface image and atomic force microscopy (AFM) topography of 3-layer IGZO 3:1:1; (b) atomic concentration (%) of each metallic cation in 3-layer combustion IGZO films with different In:Ga:Zn ratio determined by electron dispersive X-ray spectroscopy (EDS) analysis; (c) X-ray photoelectron spectroscopy (XPS) surface spectra of IGZO 3:1:1 thin films produced with and without urea (d) and In:Ga:Zn atomic percentage after argon cluster etching (0–300 s).

SEM surface images and AFM deflection (Figure 2a inset) show that smooth and uniform films are obtained regardless of processing conditions. The films roughness was determined from the AFM height profile of a 2 × 2 µm^2 area scan with rms roughness being lower than 0.3 nm for all films (Figure S3), as required for the integration in electronic devices.

Atomic percentage of metal cations was determined for combustion IGZO films by EDS analysis to determine films stoichiometry for different In:Ga:Zn ratio. Figure 2b shows that in general the films stoichiometry matches the In:Ga:Zn ratio of IGZO precursor solutions with a slight Ga deficiency for 2:1:1 and 3:1:1 films.

X-ray photoelectron spectroscopy (XPS) was performed to evaluate the structure of IGZO films produced with and without urea. Figure 2c shows high resolution spectra of the initial films' surfaces. Differential charging occurred during this measurement, thus, the spectra were charge referenced to C 1s at 284.8 eV a posteriori. The O 1s spectra of the films' surfaces are deconvoluted into three main peaks. The first component (O_I) at its lowest binding energy, 529.9 ± 0.1 eV, corresponds to M–O–M bonds [36]. The second component (O_{II}), centered at 531.3 ± 0.1 eV, is either associated with M–O–M bonds at the surface or undercoordinated oxygen [37]. The third component (O_{III}), centered at 532.3 ± 0.1 eV, is related to water and organic species adsorbed on the surface [36]. The two films have an identical C 1 s emission (not shown here), confirming a similar amount of contamination by adventitious carbon. Hence, it can be concluded that the addition of urea during the synthesis promotes the formation of M–O–M bonds at the surface. In order to address the volume of the films, XPS depth profiles were performed, given in Figure 2d. An argon cluster mode was chosen in order to induce less damage to the material than with a conventional monoatomic mode. The motivation to do depth profiles came from the cation stoichiometries observed at the surface (given in Figure 2d after 0 s of etching), which show that both films' surfaces are highly deficient in gallium, particularly the film produced without urea. For both films prepared with and without urea, the cation stoichiometries tend to match the EDS results of Figure 2b scanning the films' thicknesses towards the substrate interface. However, the cation stoichiometry of the film prepared with urea matches exactly the EDS results already after the first etching step and is maintained throughout the films' thickness. On the contrary, the gallium content of the film prepared without urea continuously increased with further etching. Two conclusions can be made from these results: first, the as-deposited surfaces of the IGZO films (with and without urea) are generally poor in gallium; second and most importantly, combustion synthesis using urea as fuel promotes film formation with a uniform cation distribution throughout the thickness, which is in line with the single exothermic peak in the DSC analysis in Figure 1.

Note that after argon cluster etching, only the first two O 1s components are observed and the third peak originating from adsorbates is no longer present (see Supporting Information Figure S4). This supports the assignment of the O 1s components made above, with the O_{II} component partially and the O_{III} component entirely related to adsorbed surface species.

3.2. Electrical Characterization of IGZO Thin Film Transistors (TFTs)

Electrical characterization of solution processed IGZO/SiO$_2$ TFTs was performed by measuring the transfer characteristics of the devices in ambient conditions in the dark to study the influence of semiconductor composition ratio and number of layers in device performance (Figure 3).

Figure 3a shows transfer characteristics of IGZO TFTs with different compositions, and IGZO TFTs with variation of layers; the statistics of the extracted parameters are represented in Figure 3b. For In-rich composition IGZO 3:1:1, the on/off current ratio (I_{On}/I_{Off}) and saturation mobility (μ_{Sat}) increases one order of magnitude compared to IGZO 2:1:1 and 1:1:1. Indium cations constitute the main element of conduction band minimum (CBM) in these amorphous structures, where potential barriers arising from random distribution of zinc and gallium cations exist. Thus, by increasing indium content, the potential barriers derived from structural randomness decrease, enhancing carrier transport. In IGZO 1:1:1 the values of I_{On}/I_{Off} and μ_{Sat} are lower as the higher gallium content helps suppress free carrier generation by forming stronger chemical bonds with oxygen when compared to zinc and indium cations. Therefore, gallium and zinc content must be tailored to guarantee amorphous films [16,38]. Still, it is relevant to notice from Figure 3a that off-current is not being significantly affected by the different IGZO compositions, being governed by the gate-to-source leakage current (I_{GS}), as expected for a TFT. However, this can also be related with the very low thickness of IGZO with 1 layer (d ≈ 14 nm), allowing for the depletion region arising from the atmosphere interaction with the IGZO back-surface to be extended through the entire IGZO films thickness [39]. The results obtained for different IGZO thickness, discussed below, shed light into this phenomenon.

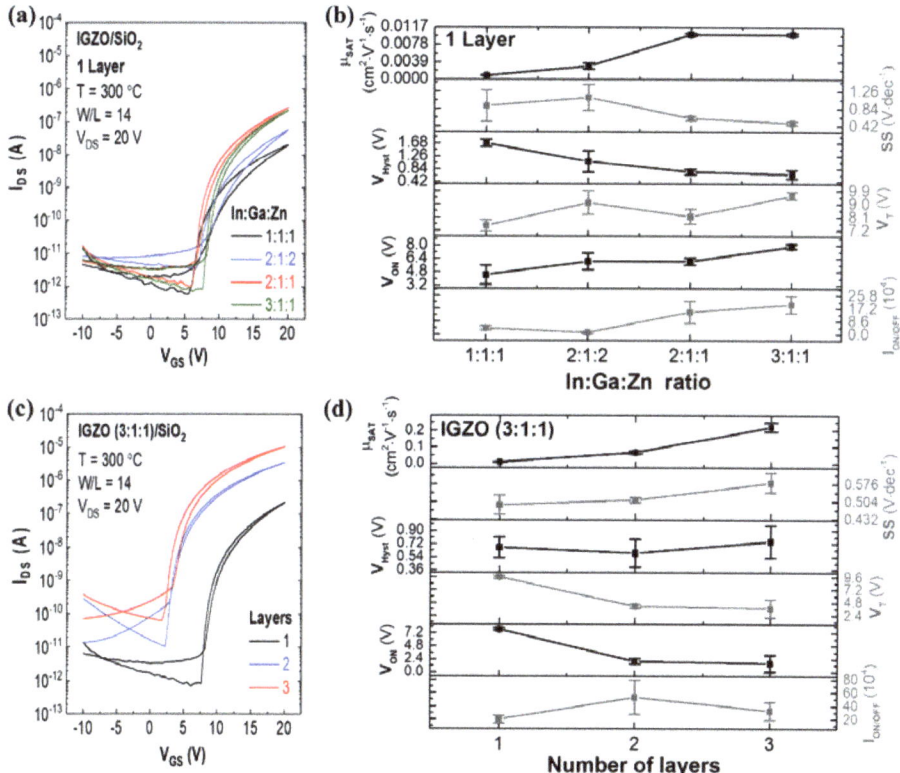

Figure 3. Transfer characteristics of IGZO thin-film transistors (TFTs) (**a**) with a different In:Ga:Zn ratio and (**b**) respective electrical parameters; (**c**) with 1-, 2- and 3-layer IGZO 3:1:1 TFTs and (**d**) their respective electrical parameters.

To understand the thickness influence, a different number of layers (1, 2 and 3) were studied in IGZO 3:1:1 TFTs (Figure 3c,d). It is evident the negative shift on V_{On} and V_T and a higher I_{Off} with the increasing number of spin coated layers, due to higher free carrier concentration (N) in the bulk of the thicker active layer, leading to charge accumulation in the semiconductor/dielectric interface; therefore the conductive channel forms at lower V_{GS} values [39]. The better electrical performance of 3-layer films can also be associated with lower porosity as with each layer deposited defects caused by gaseous products release are decreased and film densification is enhanced.

Optimized electrical performance was obtained for 3-layer IGZO 3:1:1 TFTs. As such, and to further assess the effect of urea in device performance, 3-layer TFTs were also produced using IGZO 3:1:1 precursor solution without urea (Figure S5 in Supporting information). The later devices show overall poor electrical performance with high hysteresis and low stability, thus confirming the crucial role of urea in the proper formation of IGZO at temperatures ≤300 °C.

Figure 4 shows transfer characteristics of 3-layer IGZO (3:1:1) TFTs as deposited and after 5 weeks to assess device ageing. Overall device performance (I_{On}/I_{Off}, SS and μ_{Sat}) was maintained however V_{On} and V_T show a slight negative shift over time associated to the increase of carrier density which results in the rise of oxygen vacancy concentration in the channel [40].

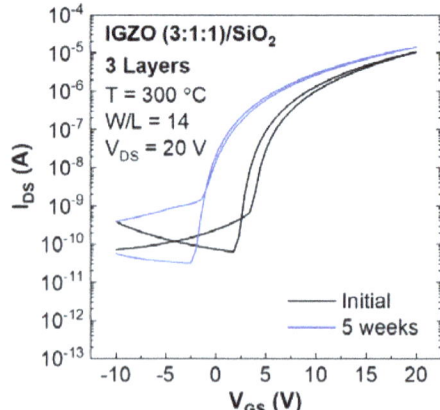

Figure 4. Transfer characteristics of a 3-layer IGZO 3:1:1 TFT as deposited and after 5 weeks.

Fully solution based TFTs were produced by combining optimized 3-layer IGZO 3:1:1 with high-κ solution process dielectric AlO_x to enable low voltage operation. Capacitance variation with frequency of Si/AlO_x/Al MIS devices is depicted in Figure S6, where C_i = 306 ± 2 nF was determined for AlO_x at 100 Hz. IGZO/AlO_x TFTs were patterned (W/L = 160/20) and passivated with chemical vapor deposited parylene.

Electrical characterization of fully solution-based 3-layer IGZO 3:1:1/AlO_x TFTs is depicted in Figure 5. TFTs transfer characteristics (Figure 5a) were obtained by varying V_{GS} from −1 to 2 V for V_{DS} of 2 V. The fully solution-based devices demonstrate enhanced performance when compared to SiO_2 non-passivated TFTs namely, I_{On}/I_{Off} = 10^6, μ_{Sat} = 3.2 cm^2 V^{-1} s^{-1}, SS = 73 mV dec^{-1}, V_{On} = 0.18, V_T = 0.63 V and I_{GS}.

The devices operational stability under positive gate bias stress (PBS) was studied in air environment by applying a constant gate voltage equivalent to electrical field of 0.5 MV·cm^{-1}, while keeping the source and drain electrodes grounded. The transfer characteristics of 3-layer IGZO (3:1:1)/AlO_x TFTs were obtained in saturation regime (V_{DS} = 2 V) at selected times during stress (Figure S7 in the Supplementary Materials) and the threshold voltage variation (ΔV_T) with time during PBS is shown in Figure 5b. The devices show a negative threshold voltage shifts under PBS, which was previously reported for IGZO TFTs using solution processed and sputtered high-κ dielectrics by our group [34,41]. The abnormal shift in V_T when applying PBS is associated to the hydrogen release from residual AlO-H bonds in the AlO_x gate dielectric and their migration to the IGZO channel. By diffusing the hydrogen atoms in the channel, a negative ΔV_T is induced through electron doping power-law time dependence [42]. Initially V_T shifts abruptly however after 30 min the devices stabilize with maximum ΔV_T = −0.22 V after 1 h of PBS.

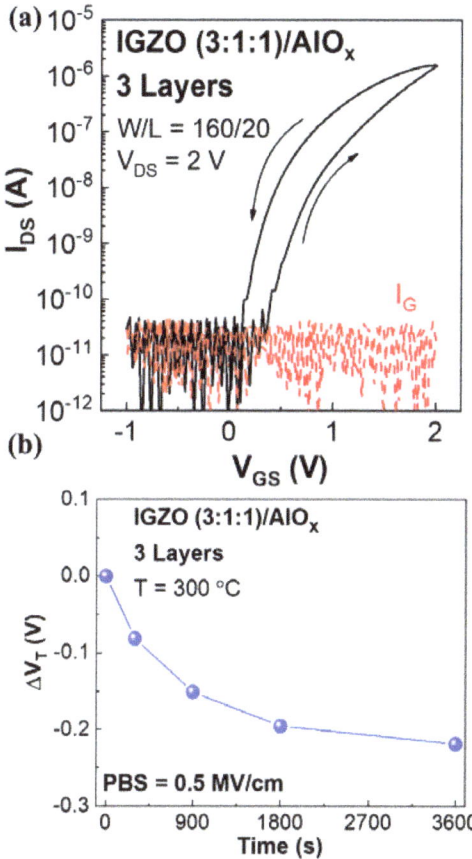

Figure 5. (a) Transfer characteristics of a fully solution-based passivated 3-layer IGZO 3:1:1/AlO$_x$ TFT; (b) threshold voltage variation (ΔV_T) under positive gate bias stress (PBS) (0.5 MV·cm^{-1}) for 1 h in air environment.

4. Conclusions

In summary, we clearly demonstrated the importance of IGZO composition and number of layers in combustion solution based IGZO TFTs. The use of urea as fuel is crucial to produce high quality IGZO films at lower temperature and assure that the precursors In:Ga:Zn ratio is maintained in the films. Indium content plays a major role to achieve enhanced electrical properties and 3-layer films show improved densification. Fully solution processed TFTs with low operation voltage were achieved with optimized 3-layer IGZO (3:1:1) active channel layer and AlO$_x$ high-κdielectric. These devices demonstrate enhanced dielectric-semiconductor interface (SS = 73 mV·dec^{-1}) and saturation mobility of 3.2 cm^2 V^{-1} s^{-1} with good stability over time. These results have been proved to be reproducible encouraging the use of fully solution based IGZO TFTs for low-cost electronic applications.

Supplementary Materials: The following are available online at http://www.mdpi.com/2079-4991/9/9/1273/s1, Table S1: Redox reactions regarding this work, Table S2: Overall oxide formation reaction considering metal nitrate reduction and urea oxidation reactions, Table S3: Calculation of oxidizing and reducing valence of reagents by the Jain method, Table S4: umber of moles of urea per mole of oxidant to ensure stoichiometry (φ = 1) of the redox reaction, Table S5: Stoichiometric overall oxide formation reactions, Figure S1: (a) FTIR spectra of IGZO solutions; (b) 3-layer IGZO thin films on Si substrates, after annealing at 300 °C for 30 min, Table S6: Characteristic absorbance peaks and associated vibrational modes of the corresponding chemical bonds for analyzed FT-IR spectra of IGZO solutions, Table S7: Viscosity of combustion IGZO precursor solutions with different In:Ga:Zn ratio, Table S8: Combustion IGZO films' thickness measured by spectroscopic ellipsometry, Table S9: Bandgap energy (Eg) of combustion IGZO thin films determined by spectroscopic ellipsometry, Figure S2: Transmittance measurements of combustion of 1-, 2- and 3-layer IGZO 3:1:1, Figure S3: AFM deflection images (2 × 2 µm2) of 1-, 2- and 3-layer IGZO thin films with different In:Ga:Zn ratio (a) 1:1:1, (b) 2:1:2, (c) 2:1:1 and d) 3:1:1, Figure S4: Deconvoluted O 1s spectra of the XPS depth profile after 0 s, 100 s and 200 s argon cluster etching, Figure S5: Transfer characteristics of 3-layer 3:1:1 IGZO TFTs produced using IGZO precursor solutions with and without urea as fuel, Figure S6: Capacitance-frequency measurements of Si/solution-based AlOx/Al MIS device, Figure S7: Transfer characteristics of 3-layer IGZO (3:1:1)/AlOx TFT when a positive gate bias stress (PBS) of 0.5 MV·cm^{-1} is applied over time.

Author Contributions: Experimental work, and fabrication of the films and devices, M.M., E.C., C.D. and M.P.; characterization of the films and devices, data analysis and manuscript preparation, M.M., E.C., C.D., J.D. and R.B.; experiments design, supervision of the work and revision of the concept, structure, and content of the different versions of the manuscript until its final form, P.B. and R.B.; funding for the fabrication and characterization facilities and reviewed the final versions of the manuscript, R.M. and E.F.

Funding: This research was funded by National Funds through FCT—Portuguese Foundation for Science and Technology, Reference UID/CTM/50025/2019 and FCT-MCTES. European Community H2020 NMP-22-2015 project 1D-NEON Grant Agreement 685758. E. Carlos acknowledges FCT-MCTES for a doctoral grant (Grant SFRH/BD/116047/2016) and IDS-FunMat-INNO project FPA2016/EIT/EIT Raw Materials Grant Agreement 15015. This work is part of two Master Thesis in Micro and Nanotechnology Engineering at FCT NOVA; i) "Composition ratio effect in IGZO using solution combustion synthesis for TFT applications" defended by M. Moreira in December 2017 and "Fully solution-based TFTs" defended by C. Dias in December 2019.

Acknowledgments: The authors would like to acknowledge J. V. Pinto and S. Pereira for XRD, A. Pimentel for TG-DSC and T. Calmeiro for AFM.

Conflicts of Interest: The authors declare no competing financial interest.

References

1. Hosono, H. Ionic amorphous oxide semiconductors: Material design, carrier transport, and device Application. *J. Non. Cryst. Solids* **2006**, *352*, 851–858. [CrossRef]
2. Nomura, K.; Ohta, H.; Takagi, A.; Kamiya, T.; Hirano, M.; Hosono, H. Room-temperature fabrication of transparent flexible thin-film transistors using amorphous oxide semiconductors. *Nature* **2004**, *432*, 488–492. [CrossRef] [PubMed]
3. Lee, W.; Choi, S.; Kim, J.; Park, S.K.; Kim, Y. Solution-free UV-based direct surface modification of oxide films for self-patterned metal-oxide thin-film transistors. *Adv. Electron. Mater.* **2019**, *1900073*, 1–6. [CrossRef]
4. Kim, G.H.; Kim, H.S.; Shin, H.S.; Du Ahn, B.; Kim, K.H.; Kim, H.J. Inkjet-Printed InGaZnO thin film transistor. *Thin Solid Films* **2009**, *517*, 4007–4010. [CrossRef]
5. Kim, D.; Koo, C.Y.; Song, K.; Jeong, Y.; Moon, J. Compositional influence on sol-gel-derived amorphous oxide semiconductor thin film transistors. *Appl. Phys. Lett.* **2009**, *95*, 1–3. [CrossRef]
6. Ahn, J.-S.; Lee, J.-J.; Hyung, G.W.; Kim, Y.K.; Yang, H. Colloidal ZnO quantum dot-based, solution-processed transparent field-effect transistors. *J. Phys. D. Appl. Phys.* **2010**, *43*, 275102. [CrossRef]
7. Kim, G.H.; Jeong, W.H.; Kim, H.J. Electrical characteristics of solutionprocessed InGaZnO thin film transistors depending on Ga concentration. *Phys. Status Solidi Appl. Mater. Sci.* **2010**, *207*, 1677–1679. [CrossRef]
8. Ong, B.S.; Li, C.; Li, Y.; Wu, Y.; Loutfy, R. Stable, solution-processed, high-mobility ZnO thin-film transistors. *J. Am. Chem. Soc.* **2007**, *129*, 2750–2751. [CrossRef]
9. Branquinho, R.; Salgueiro, D.; Santa, A.; Kiazadeh, A.; Barquinha, P.; Pereira, L.; Martins, R.; Fortunato, E. Towards environmental friendly solution-based ZTO/AlO × TFTs. *Semicond. Sci. Technol.* **2015**, *30*, 024007. [CrossRef]
10. Rim, Y.S.; Jeong, W.H.; Kim, D.L.; Lim, H.S.; Kim, K.M.; Kim, H.J. Simultaneous modification of pyrolysis and densification for low-temperature solution-processed flexible oxide thin-film transistors. *J. Mater. Chem.* **2012**, *22*, 12491. [CrossRef]
11. Kim, G.H.; Shin, H.S.; Du Ahn, B.; Kim, K.H.; Park, W.J.; Kim, H.J. Formation mechanism of solution-processed nanocrystalline InGaZnO thin film as active channel layer in thin-film transistor. *J. Electrochem. Soc.* **2009**, *156*, H7–H9. [CrossRef]

12. Kim, G.H.; Du Ahn, B.; Shin, H.S.; Jeong, W.H.; Kim, H.J.; Kim, H.J. Effect of indium composition ratio on solution-processed nanocrystalline InGaZnO thin film transistors. *Appl. Phys. Lett.* **2009**, *94*, 233501. [CrossRef]
13. Lim, J.H.; Shim, J.H.; Choi, J.H.; Joo, J.; Park, K.; Jeon, H.; Moon, M.R.; Jung, D.; Kim, H.; Lee, H.J. Solution-processed InGaZnO-based thin film transistors for printed electronics applications. *Appl. Phys. Lett.* **2009**, *95*, 93–96. [CrossRef]
14. Nayak, P.K.; Busani, T.; Elamurugu, E.; Barquinha, P.; Martins, R.; Hong, Y.; Fortunato, E. Zinc concentration dependence study of solution processed amorphous indium gallium zinc oxide thin film transistors using high-k dielectric. *Appl. Phys. Lett.* **2010**, *97*, 183504. [CrossRef]
15. Kim, Y.H.; Han, M.K.; Han, J.I.; Park, S.K. Effect of metallic composition on electrical properties of solution-processed indium-gallium-zinc-oxide thin-film transistors. *IEEE Trans. Electron Devices* **2010**, *57*, 1009–1014. [CrossRef]
16. Jeong, S.; Ha, Y.G.; Moon, J.; Facchetti, A.; Marks, T.J. Role of gallium doping in dramatically lowering amorphous-oxide processing temperatures for solution-derived indium zinc oxide thin-film transistors. *Adv. Mater.* **2010**, *22*, 1346–1350. [CrossRef] [PubMed]
17. Hwang, S.; Lee, J.H.; Woo, C.H.; Lee, J.Y.; Cho, H.K. Effect of annealing temperature on the electrical performances of solution-processed InGaZnO thin film transistors. *Thin Solid Films* **2011**, *519*, 5146–5149. [CrossRef]
18. Su, B.-Y.; Chu, S.-Y.; Juang, Y.-D.; Chen, H.-C. High-performance low-temperature solution-processed InGaZnO thin-film transistors via ultraviolet-ozone photo-annealing. *Appl. Phys. Lett.* **2013**, *102*, 192101. [CrossRef]
19. Hennek, J.W.; Smith, J.; Yan, A.; Kim, M.G.; Zhao, W.; Dravid, V.P.; Facchetti, A.; Marks, T.J. Oxygen Getter effects on microstructure and carrier transport in low temperature combustion-processed a-InXZnO (X = Ga, Sc, Y, La) transistors. *J. Am. Chem. Soc.* **2013**, *135*, 10729–10741. [CrossRef]
20. Wang, B.; Guo, P.; Zeng, L.; Yu, X.; Sil, A.; Huang, W.; Leonardi, M.J.; Zhang, X. Expeditious, scalable solution growth of metal oxide films by combustion blade coating for flexible electronics. *Proc. Natl. Acad. Sci. USA* **2019**, *116*, 9230–9238. [CrossRef]
21. Sun, Y.; Rogers, J.A. Inorganic semiconductors for flexible electronics. *Adv. Mater.* **2007**, *19*, 1897–1916. [CrossRef]
22. MacDonald, W.A. Engineered films for display technologies. *J. Mater. Chem.* **2004**, *14*, 4. [CrossRef]
23. Socratous, J.; Banger, K.K.; Vaynzof, Y.; Sadhanala, A.; Brown, A.D.; Sepe, A.; Steiner, U.; Sirringhaus, H. Electronic structure of low-temperature solution-processed amorphous metal oxide semiconductors for thin-film transistor applications. *Adv. Funct. Mater.* **2015**, *25*, 1873–1885. [CrossRef] [PubMed]
24. Kim, M.-G.; Kanatzidis, M.G.; Facchetti, A.; Marks, T.J. Low-temperature fabrication of high-performance metal oxide thin-film electronics via combustion processing SI. *Nat. Mater.* **2011**, *10*, 382–388. [CrossRef] [PubMed]
25. Epifani, M.; Melissano, E.; Pace, G.; Schioppa, M. Precursors for the combustion synthesis of metal oxides from the Sol–Gel processing of metal complexes. *J. Eur. Ceram. Soc.* **2007**, *27*, 115–123. [CrossRef]
26. Hennek, J.W.; Kim, M.-G.; Kanatzidis, M.G.; Facchetti, A.; Marks, T.J. Exploratory combustion synthesis: Amorphous indium yttrium oxide for thin-film transistors. *J. Am. Chem. Soc.* **2012**, *134*, 9593–9596. [CrossRef] [PubMed]
27. Branquinho, R.; Salgueiro, D.; Santos, L.; Barquinha, P.; Pereira, L.; Martins, R.; Fortunato, E. Aqueous combustion synthesis of aluminum oxide thin films and application as gate dielectric in GZTO solution-based TFTs. *ACS Appl. Mater. Interfaces* **2014**, *6*, 19592–19599. [CrossRef]
28. Bae, E.J.; Kang, Y.H.; Han, M.; Lee, C.; Cho, S.Y. Soluble oxide gate dielectrics prepared using the self-combustion reaction for high-performance thin-film transistors. *J. Mater. Chem. C* **2014**, *2*, 5695–5703. [CrossRef]
29. Barquinha, P.; Martins, R.; Pereira, L.; Fortunato, E. *Transparent Oxide Electronics: From Materials to Devices*; John Wiley & Sons: Hoboken, NJ, USA, 2012. [CrossRef]
30. Cheong, H.; Ogura, S.; Ushijima, H.; Yoshida, M.; Fukuda, N.; Uemura, S. Rapid preparation of solution-processed InGaZnO thin films by microwave annealing and photoirradiation. *AIP Adv.* **2015**, *5*, 067127. [CrossRef]

31. Jeong, H.; Lee, B.; Lee, Y.; Lee, J.; Yang, M.; Kang, I.; Mativenga, M.; Jang, J. Coplanar amorphous-indium-gallium-zinc-oxide thin film transistor with He plasma treated heavily doped layer. *Appl. Phys. Lett.* **2014**, *104*, 022115. [CrossRef]
32. Wang, H.; Xu, W.; Zhou, S.; Xie, F.; Xiao, Y.; Ye, L.; Chen, J.; Xu, J. Oxygen plasma assisted high performance solution-processed Al_2O × gate insulator for combustion-processed InGaZnO × thin film transistors. *J. Appl. Phys.* **2015**, *117*, 035703. [CrossRef]
33. Kagan, C.R.; Andry, P. *Thin Film Transistors*; Marcel Dekker, Inc.: New York, NY, USA, 2003.
34. Carlos, E.; Branquinho, R.; Kiazadeh, A.; Barquinha, P.; Martins, R.; Fortunato, E. UV-Mediated photochemical treatment for low-temperature oxide-based thin-film transistors. *ACS Appl. Mater. Interfaces* **2016**, *8*, 31100–31108. [CrossRef] [PubMed]
35. Salgueiro, D.; Kiazadeh, A.; Branquinho, R.; Santos, L.; Barquinha, P.; Martins, R.; Fortunato, E. Solution based zinc tin oxide TFTs: The dual role of the organic solvent. *J. Phys. D Appl. Phys.* **2017**, *50*, 065106. [CrossRef]
36. Liang, K.; Wang, Y.; Shao, S.; Luo, M.; Pecunia, V.; Shao, L.; Zhao, J.; Chen, Z.; Mo, L.; Cui, Z. High-performance metal-oxide thin-film transistors based on inkjet-printed self-confined bilayer heterojunction channels. *J. Mater. Chem. C* **2019**, *7*, 6169–6177. [CrossRef]
37. Fernandes, C.; Santa, A.; Santos, Â.; Bahubalindruni, P.; Deuermeier, J.; Martins, R.; Fortunato, E.; Barquinha, P. A sustainable approach to flexible electronics with zinc-tin oxide thin-film transistors. *Adv. Electron. Mater.* **2018**, *4*, 1–10. [CrossRef]
38. Olziersky, A.; Barquinha, P.; Vilà, A.; Magaña, C.; Fortunato, E.; Morante, J.R.; Martins, R. Role of Ga_2O_3–In_2O_3–ZnO channel composition on the electrical performance of thin-film transistors. *Mater. Chem. Phys.* **2011**, *131*, 512–518. [CrossRef]
39. Barquinha, P.; Pimentel, A.; Marques, A.; Pereira, L.; Martins, R.; Fortunato, E. Influence of the semiconductor thickness on the electrical properties of transparent TFTs based on indium zinc oxide. *J. Non-Cryst. Solids* **2006**, *352*, 1749–1752. [CrossRef]
40. Song, Y.; Katsman, A.; Butcher, A.L.; Paine, D.C.; Zaslavsky, A. Temporal and voltage stress stability of high performance indium-zinc-oxide thin film transistors. *Solid. State. Electron.* **2017**, *136*, 43–50. [CrossRef]
41. Martins, J.; Bahubalindruni, P.; Rovisco, A.; Kiazadeh, A.; Martins, R.; Fortunato, E.; Barquinha, P.; Martins, J.; Bahubalindruni, P.; Rovisco, A.; et al. Bias stress and temperature impact on InGaZnO TFTs and circuits. *Materials* **2017**, *10*, 680. [CrossRef] [PubMed]
42. Chang, Y.-H.; Yu, M.-J.; Lin, R.-P.; Hsu, C.-P.; Hou, T.-H. Abnormal positive bias stress instability of In–Ga–Zn–O thin-film transistors with low-temperature Al_2O_3 gate dielectric. *Appl. Phys. Lett.* **2016**, *108*, 033502. [CrossRef]

© 2019 by the authors. Licensee MDPI, Basel, Switzerland. This article is an open access article distributed under the terms and conditions of the Creative Commons Attribution (CC BY) license (http://creativecommons.org/licenses/by/4.0/).

MDPI
St. Alban-Anlage 66
4052 Basel
Switzerland
Tel. +41 61 683 77 34
Fax +41 61 302 89 18
www.mdpi.com

Nanomaterials Editorial Office
E-mail: nanomaterials@mdpi.com
www.mdpi.com/journal/nanomaterials

www.ingramcontent.com/pod-product-compliance
Lightning Source LLC
LaVergne TN
LVHW071954080526
838202LV00064B/6750